Approaches to Handling Environmental Problems in the Mining and Metallurgical Regions

T0137947

NATO Science Series

A Series presenting the results of scientific meetings supported under the NATO Science Programme.

The Series is published by IOS Press, Amsterdam, and Kluwer Academic Publishers in conjunction with the NATO Scientific Affairs Division

Sub-Series

I. **Life and Behavioural Sciences**	IOS Press
II. **Mathematics, Physics and Chemistry**	Kluwer Academic Publishers
III. **Computer and Systems Science**	IOS Press
IV. **Earth and Environmental Sciences**	Kluwer Academic Publishers
V. **Science and Technology Policy**	IOS Press

The NATO Science Series continues the series of books published formerly as the NATO ASI Series.

The NATO Science Programme offers support for collaboration in civil science between scientists of countries of the Euro-Atlantic Partnership Council. The types of scientific meeting generally supported are "Advanced Study Institutes" and "Advanced Research Workshops", although other types of meeting are supported from time to time. The NATO Science Series collects together the results of these meetings. The meetings are co-organized bij scientists from NATO countries and scientists from NATO's Partner countries – countries of the CIS and Central and Eastern Europe.

Advanced Study Institutes are high-level tutorial courses offering in-depth study of latest advances in a field.
Advanced Research Workshops are expert meetings aimed at critical assessment of a field, and identification of directions for future action.

As a consequence of the restructuring of the NATO Science Programme in 1999, the NATO Science Series has been re-organised and there are currently five sub-series as noted above. Please consult the following web sites for information on previous volumes published in the Series, as well as details of earlier sub-series.

http://www.nato.int/science
http://www.wkap.nl
http://www.iospress.nl
http://www.wtv-books.de/nato-pco.htm

Approaches to Handling Environmental Problems in the Mining and Metallurgical Regions

edited by

Walter Leal Filho
Technical University Hamburg-Harburg Technology GmbH (TuTech),
Environmental Technology,
Hamburg, Germany

and

Irina Butorina
Priazovsky State Technical University,
Mariupol, Ukraine

Kluwer Academic Publishers

Dordrecht / Boston / London

Published in cooperation with NATO Scientific Affairs Division

Proceedings of the NATO Advanced Research Workshop on
Approaches to Handling Environmental Problems in the Mining and Metallurgical
Regions of NIS Countries
Mariupol, Ukraine
5–7 September 2002

A C.I.P. Catalogue record for this book is available from the Library of Congress.

ISBN 1-4020-1322-1 (HB)
ISBN 1-4020-1323-X (PB)

Published by Kluwer Academic Publishers,
P.O. Box 17, 3300 AA Dordrecht, The Netherlands.

Sold and distributed in North, Central and South America
by Kluwer Academic Publishers,
101 Philip Drive, Norwell, MA 02061, U.S.A.

In all other countries, sold and distributed
by Kluwer Academic Publishers,
P.O. Box 322, 3300 AH Dordrecht, The Netherlands.

Printed on acid-free paper

Printed in the Netherlands.

TABLE OF CONTENTS

vi

Preface

This book is one of the outcomes of the NATO Advanced Research Workshop "Approaches to handling environmental problems in the mining and metallurgical regions of NIS countries" held in Mariupol, Ukraine on 5-7 September 2002.

It include papers written by some of the leading specialists in the field of mining and metallurgy, and by environment specialists who are active in this sector. Readers will notice that some common environmental problems seen in the mining and metallurgical industries are described and that their influence on the health of the population are discussed. Examples of best practice in the field are given both from EU countries and from Central and Eastern European nations, especially from the Newly Independent States (NIS).

Some of the latest technologies involved in the elimination of hazardous emissions, in sewage treatment and the handling of wastes in the metallurgical and mining industries are presented and we hope that they may open the way for more West-East, East-West and East-East technology and know-how exchange.

In preparing this book, thanks are due to Marina Butorina, Linda Döring and Olaf Gramkow for their competent advice in respect of translations, lay-out and handling of the texts. We also grateful to NATO´s Scientific Affairs Division for the support with the workshop, whose benefits are already being felt in both Mariupol and elsewhere in Eastern Europe.

Prof. Walter Leal Filho and Dr. Irina Butorina

CHAPTER 1

Environmental problems related to mining in the industrial regions of Europe

Prof. Walter Leal Filho

TUHH-TuTech
Kasernenstr. 12
D-21073 Hamburg
Germany

Summary

This paper outlines some of the environmental problems related to mining in Europe and some of the work performed in Europe and elsewhere, with a view to minimising them and to moving towards more sustainable mining practice.

Introduction: the importance of mining

The mining industry has been a key to the development of civilisation, underpinning the Iron Age and Bronze Age, the industrial revolution and the infrastructure of today's information age (Mbendi 2002). In 2001, the mining industry produced over 6 billion tons of raw product, valued at several trillion dollars. Downstream beneficiation and minerals processing of these raw materials adds further value as raw materials and products are created to serve all aspects of industry and commerce worldwide.

The last decade of the twentieth century saw the creation of mega-commodity corporations that increasingly moved downstream into the beneficiation area, leaving exploration for new mineral deposits increasingly to small junior mining companies. Application of new technology has led to productivity gains across the value chain.

Apart from Antarctica (which has a treaty in place preventing short to medium term exploitation and exploration for minerals), mining takes place in all of the world's continents. Traditional mining countries such as the US, China, Australia, South Africa and Chile dominate the global mining scene. These countries have become the traditional leaders in mining and exploration methods and technology. Exploration and development funding has changed over the past few years, with emphasis shifting to areas that have been poorly explored or have had poor access for reasons of politics, infrastructure or legislation. Gold, base metal, diamonds and platinum group elements (PGE's) are the more important commodities explored for and developed globally. Nonetheless, the following categories of mining are all seen:

- metal mining (including gold and iron)
- coal mining
- salt and potash mining
- minerals mining (fluospar, baryte, calcspar, graphite, clay, slate, limestone, asphalt, gypsum, etc)

W. Leal Filho and I. Butorina (eds.),
Approaches to Handling Environmental Problems in the Mining and Metallurgical Regions, 7–16.
© 2003 *Kluwer Academic Publishers. Printed in the Netherlands.*

Europe is not a major mining centre. However, it has several established base metal mines in Scandinavia, Ireland, the Iberian Peninsula and in the Eastern region. Major companies include Boliden and Outokumpu.

Base metals and gold are produced in Ireland, Spain, Portugal, Romania, Turkey, Sweden and Finland. Cyprus is well known for its base metal deposits on Troodos. Turkey has great potential for base metal and gold deposits; and is an established chromite producer. Coal is a major European product, with Germany and Poland being major producers.

Poland, once Europe's major producer, is suffering from declining production. All coal operations are owned and controlled by the Ministry of Economy, which has reported a decline in production from 117 Mt in 1998 to 101 Mt in 2000. However, Poland's exports have increased slightly, up to 24Mt in 2000 from 21 Mt in 1999.

Germany's coal production continues to decline from 41Mt in 1998 to 33Mt in 2000. Ruhrkohle (RAG) are responsible for most of Germany's production, as well as having interests in Latin America. Germany is the world's largest lignite producer. Greece is also a major lignite producer. Public Power Corporation produces 58 million tons of the total 60.4 million tons produced annually. The United Kingdom produced 41 million tons in 1998 and Spain 26 million tons of coal in 1998.

In some countries, mining of what are regarded as hazardous or dangerous materials, such as uranium, still takes place. In Germany, although several uranium deposits were discovered and explored in the highlands, no commercial uranium mining has developed. Test mines existed in Ellweiler (Rhineland-Palatinate), Baden-Baden/Gernsbach in the northern part of the Black Forest, Menzenschwand in the southern part of the Black Forest, Mähring and Poppenreuth in Northern Bavaria, and Großschloppen in the Fichtel Gebirge. The only uranium mill was in operation from 1961 to 1989 at Ellweiler. It produced a total of around 700 tons of uranium, mainly from Menzenschwand ores. After protests by environmental activists, it was shut down in 1989 for exceeding radiation release limits from the associated mill tailings dump. In Mähring, heap leaching was continued for some period of time after the shutdown of the test mine. At the end of the eighties, all uranium exploration and mining activities in Germany were discontinued due to the low market price of uranium (Diehl, 1995).

In the Ukraine, uranium is being mined in the Ingul'skii and Vatutinskii mines near Kirovograd. The ore is processed in the Zholtiye Vody and Dneprodzerzhinsk mills. There are few details available on Ukrainian uranium production. The annual production for 1992 was estimated at 1000 tons. In April 1995, the Ukrainian government approved a nuclear fuel industry plan, scheduling a threefold increase of uranium production by the year 2003. Along with Spain, the Ukraine is the only country where the production of uranium has visibly increased

In Russia, uranium is at present only being mined at the Streltsovsk deposit in the eastern Transbaikal district in Eastern Siberia. The current annual production is estimated at 3000 tons. In the European part of Russia, a small uranium deposit in the Onezsk district in Karelia is known, but has not yet been mined. The deposits in the Stavropol district and in the Northern Caucasus Mountains have been exhausted.

France is by far the largest uranium producer in Western Europe. In 1988, production reached a peak of 3394 tons; this allowed France to meet half of its reactor demand from domestic sources.

Environmental problems related to mining

Mining, as an economic activity, has made deep impressions on many people living close to mining sites (Rowland 1999) or even far from them. It also has a clear gender bias towards men (Lahiri-Dutt 2002).

Despite its economic success in most areas (Mining Journal 2002), mining has also had a history associated with environmental mismanagement. A great number of wetlands, lakes, streams and forests have been damaged or completely destroyed by some mining operations, which helps to explain why such activities are unpopular in many parts of the world. According to the organisation "Global Forest Watch", an international data and mapping network that combines on-the-ground knowledge with digital technology to provide accurate information about the world's forests, mining is one of the main activities leading to the destruction of forests.

Mining operations, whether or not related to the so-called "hazardous products", have always been closely related to environmental problems. The European hot spots such as the Ruhr-region in Germany or the black triangle in Poland to name but a few, have had a history of environmental problems related to mining operations. Among the environmental problems related to mining, the following may be mentioned:

- pollution (air, water, soil)
- habitats degradation or destruction of ecosystems
- contamination of rivers and groundwater
- noise

Other problems may also occur depending on the nature of the mining activity and the specific issues related to the product being mined (i.e. radioactive minerals). It is therefore expected that a wide range of environmental issues are associated with mining. An overview of some of such issues is seen in Table 1, which lists some of the many environmental issues that are associated with mining.

Table 1- Some environmental issues associated with mining

Environmental degradation
Environmental impact assessment
Environmental management systems
Environmental awareness and training
Environmental monitoring and performance
Tailings containment
Rehabilitation and revegetation
Environmental auditing
Hazardous materials management, storage and disposal
Noise, vibration and airblast control
Landform design for rehabilitation
Cyanide management
Dust control
Water management
Environmental risk
Contamination of sites/ pollution
Cleaner production

It can be seen that mining operations are not only complex per se, but that they are also related to various environmental issues. In the mining region of Duisburg, for example, the following mining plants are seen:

I. ThyssenKrupp Stahl AG (August Thyssen Hütte)

Schwelgern works: (blast furnaces)
Bruckhausen works: (blast furnaces)
Beeckerwerth works: (steel plant, rolling mills)
Meiderich works: (steel)
Hüttenheim works: (heavy plate mill)

II. ISPAT
Ruhrort plant (steel plant, rolling mills; blast furnaces), demolished 1995:
Recycling und Roheisen GmbH
Duisburger Kuferhütte (blast furnaces, sintering plant, power station)
IV. HKM (Hüttenwerk Krupp-Mannesmann)
Huckinger Hütte: (blast furnaces, coke works, sintering plant, steel plant)
V. M.I.M. Hüttenwerke Duisburg GmbH (MHD)
Metallhütte Duisburg (lead-zinc-production with an imperial smelter, sintering plant)

In addition to the ThyssenKrupp Stahl AG in Germany, the list of major mining companies include Alcan, Comalco, Rio Tinto, De Beers, BHP Billiton, Anglo Platinum, Norilsk Nickel, Pasminko and others.

Remote sensing for mining: a practical project

Based on discussions between mining companies and consultants who recognised the potential for combining prospecting and monitoring techniques, a project titled "New technologies for mineral exploration and surveillance of environmental impacts of mining operations based on remote sensing and multi-data set analysis" was initiated, funded by the Programme "Industrial and Materials Technologies" (BRITE-EURAM/CRAFT/SMT) of the European Commission. The project had two main objectives. Firstly, the partners wanted to find ways to use existing data and new remote sensing data in the search for hidden mineral deposits. Secondly, they wanted to extend the use of remote sensing for the assessment, surveillance and monitoring of the environmental impact of mining operations. The main technical challenge was to make the system sufficiently flexible to accommodate the requirements of all the potential users.

Four organisations with experience in mining, geodata processing and remote sensing have collaborated in using various spatial analysis methods, including a Geographical Information System (GIS), to correlate a wide variety of geographical, geological, geophysical, meteorological and industrial data of interest to the mining industry. Much of the information was gathered by remote sensing from aircraft and satellites. By focusing on three sites in Spain and Germany, the partners have demonstrated the potential of dedicated digital analysis procedures for mineral prospecting and for environmental monitoring of mining activities.

The project started with research focusing on three sites: one in central Spain, one in southern Spain and one in the Erzgebirge of eastern Germany.

• The Pyrite Belt covers more than 8,000 km^2 and stretches from south-western Spain into Portugal. The deposits include copper, zinc, lead, tin and manganese. It has been mined for more than 4,000 years and several companies are still extracting there. This long history of exploitation has left a legacy of spoil heaps which cause serious environmental problems. The Mediterranean climate is ideal for evaluating the ability of remote sensing and GIS techniques to detect and monitor contamination from mining activities. The project concentrated on the central part of the Belt in the Huelva province of Spain, one of the few places in Europe where there is active mining.

• The Alcudia Anticline is located to the east of the Pyrite Belt, south of Almadén. It has been intensively mined for lead, zinc and silver since Roman times. There is no mining there now, but more than 500 small mines were extracting only a century ago. It is not a promising area for exploration, but is ideal for using remote sensing to investigate contamination from old mines in numerous small places. Roman contamination is especially interesting since it is already covered by soil and difficult to map.

• The Eastern Erzgebirge, on the border between Germany and the Czech Republic, is one of the most extensively mined areas in Europe. Deposits of tin, tungsten, uranium, gold, silver and zinc have been exploited for more than 1,000 years. The region is characterised by a high population density and large forests. Environmental damage caused by mining and related heavy industry is widespread and can be severe.

The first step was to collect available information for each of the three areas and integrate it in the GIS database. For the Erzgebirge, for example, there were 85 'layers' of information, including 30 sets of remote sensing data. Information included such things as satellite images and radar maps, aerial photographs, soil geochemistry, gravity surveys, aeromagnetic data, geological maps, topographical maps, sources of pollution, meteorological records, and so on. Similar databases were compiled for the two Spanish areas. New surveys were undertaken in some areas, where the need arose.

Once all these data had been converted into GIS format, researchers were able to relate numerous different measurements to each other at each point on the ground. Complex statistical methods were devised to restore gaps in irregular and incomplete data sets and to explore the relationships between different types of data and how they relate to geological and environmental features of interest. In this very productive collaboration, the partners used the databases to pursue 15 different lines of inquiry.

The group attempted to identify hidden mineral reserves by using modelling of geophysical data, extrapolating from surface data and by simulating the local geology. Possible new target areas have been identified in the Erzgebirge. Further detailed research along these lines promises to be fruitful. The most remarkable results were achieved in environmental monitoring. By sophisticated analysis of data in the Erzgebirge, the researchers showed that the loss of forest by acid rain which has occurred in the last few decades is not directly linked to the presence of ore mining activities. Damage to the forests and associated poor air quality is, rather, governed by topographical and climatological factors and the presence of lignite-fired power stations and associated heavy industry.

Work on a site in the Alcudia Anticline demonstrated the potential of remote sensing to detect 2,000-year-old mining contamination through its effect on vegetation, even where the contamination is now covered by half a metre of new soil. The method, however, needs to be modified for other sites where soil and vegetation are different. Another important finding was that ancient lead and silver extraction does not appear to have had a harmful effect on recent vegetation now covering the site. The partners believe that using GIS and remote sensing techniques to identify ancient contamination sites saves about a third of the time and manpower required by traditional survey methods.

One advantage of the GIS approach is that smaller countries and companies can have access to data that has historically been the preserve of large multinational companies with well-resourced exploration divisions. Small mining companies should now be in a better position to compete. Although the project was designed for the mining industry, especially for the exploitation of resources and resulting environmental impact, it is hoped that the principle of correlating many different sets of information will in future be applied to a wide range of industries that need to handle specialised data. The mathematical tools developed in the project should have a wide market, not only in the mining field.

Towards sustainable mining in Europe

As known deposits of valuable minerals are consumed, the mining industry needs to employ increasingly sophisticated techniques in its search for new reserves. At the same time, mining companies are becoming more sensitive to environmental issues and are looking for new tools to help them monitor and reduce the damaging effects of their operations.

In Europe, exponents of best environmental practice in mining share the following attributes:

– an environmental ethos where environmental and business excellence go hand in hand;
– continued motivation and willingness to improve environmental management systems and performance, including training and awareness-raising initiatives;
– a clear understanding of their environmental impact and responsibilities;
– the involvement of employees and senior management on an equal basis;
– a positive attitude towards environmental issues, which are seen as a chance or opportunity and not as a threat.

Internationally, the NGO "Mining Impact Coalition" (MIC) formed a global coalition to educate people, conduct research and facilitate communication about the:

– social, economic and environmental impacts of unsafe mining
– sustainable use of the world's mineral resources

There are also many specialised companies providing groundwater hydrology, geochemistry, and remediation services, tailor-made to the mining industry.

In addition to standards in health and safety (Hendryson 1980, Chugh and Wangler 1998), methods to optimise mining and minimise their environmental damage, yet at the same time increasing the quality of the product, have been pursued worldwide (Postle 1987). Table 2, for example, provides a complete break up of Adaro Envirocoal. It can be seen that this eco-friendly coal contains 1 per cent ash and 0.1 per cent sulphur, making it one of the purest forms of coal available on the market. In addition, a wide range of equipment such as:

- gas detectors
- flame detectors
- industrial water filters
- cooling towers,
- wastewater re-use and
- recycling plants

are available for the mining and industrial industries. Moreover, services related to the granting of permission for, reclamation and development of mine sites for mining companies, public agencies, developers, and institutions are also provided. Last but not least, some institutions also provide research and education on the impacts of unsafe mining.

It should be pointed out that environmental protection in mining is guided by the concept that mine sites can become attractive and productive landscapes by integrating into an organised plan:

- the unique qualities of the geologic formations;
- the specific earth handling procedures of the mining operation; and
- the particular requirements of end uses.

The undertaking of environmental impact assessments, feasibility studies, pilot plant evaluations, engineering design and computer simulation modelling in the design and management of complex mining systems, may reduce the negative environmental risks in mining. The integration of the following may help to lower the risks of problems in the short and long term:

Sound Science: enthusiastic and highly experienced engineering and science professionals employing sound scientific practices to produce reliable results;

Smart Environmental Strategies: quick, cost-conscious strategies towards efficiently protecting environmental resources and meeting regulatory requirements and

Effective Resource Management: effective leadership and follow-through in achieving environmental management objectives.

Table 2- Composition of Adaro Envirocoal

Moisture (a.r.b) %	25.0	Ash Fusion Temperature °C (Reducing Atmosphere)	
Moisture (a.d.b.) %	16.5	Deformation	1200
Ash (a.d.b) %	1.0	Hemisphere	1260
Volatile Matter (a.d.b.) %	41.5	Flow	1340
Fixed Carbon (a.d.b.)%	41.0	Typical Ash Composition % (Dry Basis)	
Calorific Value (a.d.b.) kcal/kg	5900	Silicon (SiO_3)	35.0
Calorific Value (g.a.r.) kcal/kg	5250	Aluminium(Al_2O_3)	25.0
Total Sulphur %	0.1	Iron (Fe_2O_3)	15.0
H.G.I.	48	Calcium (CaO)	10.0
Ultimate Analysis % (Dry Ash Free Basis)		Magnesium (MgO)	4.0
Carbon	74.3	Sodium (Na_2O_3)	1.0
Hydrogen	5.6	Potassium (K_2O_5)	1.0
Nitrogen	1.1	Phosphorum (P_2O_5)	0.1
Oxygen	19.0	Titanium (TiO_2)	0.8
Sulphur	0.1	Manganese (Mn_3O_4)	0.1
Other Properties % (a.d.b.)		Sulphur (SO_3)	8.0
Chlorine	0.01		
Phosphorus	0.0005		
Carbonate	0.09		
Fluorine	0.005		

Figure 1 illustrates a site which has been mined, but which shows no sign of it.

Figure 1- Former mining site in Mystic Bay, state of Indiana, United States.
(Copyright by Bauer Ford Reclamation Design)

The example from Mystic Bay shows that properly planned pits and quarries can increase the value of mined-out land with minimum costs to or disruption of the mining operation.

A further tool towards sustainable mining practice in Europe is legislation and, in this context, examples from other parts of the world can be useful to Europe. In the United States for example, under existing federal regulations, mining companies are required to purchase bonds or cash equivalents to cover the costs of environmental clean-ups from mining operations on public lands.

The up-front bond purchases ensure that toxic pollution will be cleaned up and mutilated landscapes reclaimed when a company is unable or unwilling to pay. These financial guarantees are needed because the mining industry has left a trail of toxic contamination: 500,000 abandoned mines across the county; the pollution of 40 percent of Western watersheds; and nearly 30 mine sites currently on the EPA's Superfund National Priority List of the most contaminated sites in the nation.

Typically, a large mine site will disturb more than 10,000 acres of land. At many mines, acid drainage can result in the leaching of toxic and carcinogenic contaminants such as arsenic into water sources. The costs of addressing this pollution can amount to over one hundred million dollars per mine. Mining companies have been underestimating the cost of the clean-ups, and have failed to provide adequate guarantees to ensure that the mines will be cleaned up. Kuipers estimates that mining companies in the U.S. have a potential gap of $10 billion in clean-up coverage. If the mining companies prove unable to pay for the clean-ups, taxpayers will have to pay.

A further example of what can be done to pursue the goal of sustainable mining comes from Australia. The Best Practice in Environmental Management in Mining program (BPEM) produces a set of booklets which are useful reference materials to encourage, assist and lead all sectors of the resources industry towards achieving sustainable development.

Since 1994, 21 booklet titles have been published covering many aspects of mining. Each booklet focuses on a key environmental issue in mining, describing best practice principles and their application. Case studies are drawn from operating Australian mines and references given for further reading. The booklets are not designed to be comprehensive technical manuals, but instead attempt to give practical advice on understanding and responding to environmental problems.

Whilst the booklets are written with a focus on Australian and regional operations, their content is applicable worldwide. The first nine titles are also available in Mandarin and Spanish. The series has won international acclaim and is highly regarded as a practical guide to best practice mining operations. In 1997, the World Bank acclaimed the series as 'The current benchmark for the industry'. The BPEM program will continue to be developed with revisions to some of the older more popular titles and new titles being prepared. Some of the topics currently being considered include: Mine Decommissioning, Non-dust Atmospheric Emissions and Energy Efficiency, and Greenhouse Gas Reduction.

All in all, mining is still an important economic activity and its environmental impacts need to be constantly monitored and controlled (Struthers, 2002) so that the community is not left with a legacy of environmental damage. Today's technologies are quite advanced and may allow us to exercise the necessary control and ensure that the economic dependence of certain regions on mining does not imply a deterioration in their quality of life.

References

Chugh, Y.P., Wangler, G. (1998) "Total Health and Safety Management in Coal Mines" Proceedings of the 5th Int. Conference on Environmental Issues and Waste Management in Energy and Mineral Production, eds. G. Pasamehmetoglu and A. Ozgenoglu, May 1988, Ankara, Turkey.

Diehl, P. (1995) Uranium Production in Europe. Published by WISE News Communique in September 1995.

16

Hendryson, E. (1980) Health and safety aspects of mining: a bibliography. Medical Center Library, University of New Mexico.

Lahiri-Dutt, K. (2002) Mainstreaming gender in mining and sustainable economic development. In *Mining Environmental Management, May 2002.*

Mbendi (2002) Mining Information. Mbendi, Johannesburg.

Mining Journal (2002) Rio Tinto Rides Storm. In *Mining Journal, Volume 339. No.8695*, 2002.

Postle, J.T. (1987) Mining: An Overview. Address given at the 1987 Small Scale Mining Seminar, Thunder Bay, Ontario, April 23-29, 1987.

Rowland, D. J. (1999) Whites Crossing. Carlisle, PA: Sunset Publications

Sevin, H. and Lei, D. (1998) The Problem of Production Planning in Open Pit Mines. In *Information Systems and Operational Research*, Ottawa, Ontario, Canada, Vol. 36, No. 1, March 1998.

Struthers, S. (2002) Recycling and Total Project Development - the key to sustainable mining. In *Mining Environmental Management, May 2002.*

CHAPTER 2

Environmental problems in the mining and metallurgical Donbass region

Dr. Svyatoslav Kurulenko

State Department of the Environment
and Natural Resources
Ministry of Natural Resources, Donetsk Region
Pushkin's Parkway, 13
83000, Donetsk
Ukraine

Summary

This chapter presents an overview of the environmental problems associated with the mining and metallurgical sector in the Dombass region, Ukraine.

Introduction

As is known, the United Nations Organization has proclaimed as one of the major tasks of the world community for the 21st century the solving of the problem of the sustainable development of the economy and such ways of life that lead to a balance of environmental management and capacity of nature. The necessity for this is caused by an increasing load of unbalanced economic activities on all aspects of the environment.

The prediction of I.V. Vernadskij, the First President of the Ukrainian Academy of Sciences, that mankind will become a major geological force rendering influence on a planetary scale on the biosphere, has come true. And only if people comprehend the fragility of the biosphere will harmony between mankind and nature be possible.

First of all, people who live and operate in technogenically overloaded regions, including us, the inhabitants of the Donetsk Region, should realize this.

Historical development has caused our region to become the most polluted region in the Ukraine. A stream of harmful emissions to the atmosphere, sewage into reservoirs and industrial and consumer waste products pose the greatest problems here.

In 2000 hazardous emissions in this region amounted to 40 percent of the total emissions in the Ukraine. 1.6 million tons of hazardous substances from stationary sources and more than 200 thousand tons from vehicles were emitted to the atmosphere. As a result 70 tons of atmosphere pollutants per square kilometre are emitted. This figure is 10 times higher than the same parameter in the Ukraine as a whole.

In 2000 the dumping of polluted and insufficiently purified sewage amounted to 943 million m^3 or 30 percent of the total dumping of polluted sewage in the Ukraine.

W. Leal Filho and I. Butorina (eds.),
Approaches to Handling Environmental Problems in the Mining and Metallurgical Regions, 17–22.
© 2003 *Kluwer Academic Publishers. Printed in the Netherlands.*

The amount of accumulated industrial wastes has reached a total of 4 billion tons and municipal wastes 400 million m³. Waste production represented 30 percent of the total for the Ukraine in 2000.

The dumps and stores of waste products, open casts and soil refuse dumps, used industrial sites demanding soil recultivation already occupy about 2 percent of the territory of our region. And this process is continuing.

As a result, the industrial cities of the region exceed the maximum permitted concentration of certain hazardous substances (dust, nitric dioxide, ammonia, phenol, fluoric hydrogen, sulphur oxides, hydrogen sulphide, formaldehyde, benzapilen etc.) in the atmosphere by 1.3 to 5.8 times. These parameters were fixed in the cities of Mariupols, Donetsk, Makeyevka, Yenakievo, Gorlovka, Kramatorsk, Slavyansk, and Dzerzhinsk where periodic measurements of the atmospheric concentration of hazardous substances were conducted.

The greatest influence on atmospheric pollution must be attributed to thermal power stations, metallurgical, coke and chemical factories and the coal industry. In 2000 a total of 1284 local plants emitted more than 25 tons of hazardous substances to the atmosphere.

The main polluters of water objects in our region are metallurgical, coke and chemical plants, the coal industry, and housing and communal services.

Almost all the rivers in the region show a high concentration of salts. One of the reasons for this is the dumping of highly mineralised mine waters that bring more than 1 million tons of salts to rivers. In addition, 17 thousand tons of weighted substances, 16 thousand tons of nitrates, 6 thousand tons of hazardous organic substances, 1.5 thousand tons of nitrogen ammonia and a number of other hazardous substances were dumped together with the sewage from various branches of industry into the waters of the region.

Water basins which are sources of the centralized drinking water supply, do not meet normative requirements because of their increased mineralization and rigidity. The state of normative purification and drainage of municipal sewage is not satisfactory in the region. The norms for purification and drainage of sewage are not fulfilled satisfactorily in this region.

The majority of the rivers are silted up. They have been classified as dirty or extremely dirty waters. Both the drainage of polluted sewage, including storing drains, into freshwater sources and the long-term accumulation of polluting substances in ground sedimentation, are some of the causes of this process.

In 2000 there was still a high level of polluting substances in the coastal waters of the Sea of Azov. The mineral oil content was 5.6 times higher than maximum allowed concentrations (MAC) of nitrates, detergents and phenols. It is necessary to note that in recent years the dumping of polluted sewage into the waters of the region has stabilized. However, it still remains extremely high.

The ecological situation in the Donetsk region is further complicated by the extremely large amount of accumulated waste products, including toxic wastes. The majority of facilities storing industrial and solid municipal wastes do not meet ecological and sanitary-and-hygienic requirements. They do not prevent toxic elements from entering into surface and underground waters, soil or the atmosphere and, besides, pollute biological resources. This process also continues.

The main sources of large-tonnage waste are the coal and steel industry, power stations and the mining of non-metallic minerals. The sources of formation of toxic and epidemiological dangerous wastes are the chemical and coke industry, mechanical engineering and metal working, black and nonferrous metallurgy, agriculture and municipal wastes.

585 million tons of toxic industrial wastes and about 100 million tons of environmentally dangerous solid household wastes were accumulated in the region according to data gathered in the year in 2000. The formation of industrial wastes has reached 52 million tons in the region, 24.6 million tons of which being toxic wastes. Besides, about 1.5 million tons of household waste products have been identified in the same year.

The rate of neutralization and recycling of industrial wastes at 10-12 percent of the total amount of waste produced is completely insufficient. Processing and recycling of accumulated household wastes, except for partial extraction of metals, is practically non-existent.

The intensive development of minerals and their processing negatively influences the geological and environmental situation, makes exogenic geological processes active, changes the physicomechanical properties of soil, and the structure of underground and surface water. In connection with the closure of unpromising collieries, it has become necessary to provide a complex estimation of the challenge which the ecological situation poses, to develop and implement actions which will minimise the negative consequences arising from the closure of the collieries, this being one more point to add to the list of existing environmental problems caused by mining activities.

One of the most pressing questions related to the provision of rational environmental management in the region is the improvement of use, protection and restoration of fertility of the land. At 65 percent the rate of ploughed land is high. 70 percent of arable lands are eroded. The amount of humus in the soil is being reduced from year to year. 25 thousand hectares of agricultural land are ruined as a result of industrial activity. 4.5 thousand hectares are wasted and should be recultivated.

The number of all kinds of wild animals is still decreasing in the region. Fish productivity is declining in inland reservoirs and the Sea of Azov. The flora of this area is rich and varied. However, it reacts sensitively to the massive impact of technology. This has led to a change of population structures and as a result 34 species of flora have disappeared over the past few decades.

7.6 percent of the land is covered by forest, this being much lower than in other regions of the Ukraine. The condition of the woods is satisfactory as a whole, with the exception of the woods belonging to the agricultural enterprises.

The nature reserve area has covered 2.86 percent of the territory in 2000. Four regional landscape parks – "Donetsk Kryazh", "Klepan Byk", "Meotida" and "Polovetskaya Steppe" created in 2000 by the decision of the Donetsk Regional Administration were added to the number of institutions of nation-wide value, such as the Donetsk Botanical Garden of the National Academy of Ukraine, the National Natural Park "Saint Rocks" and the Ukrainian Steppe Natural Reserve. However, the percentage of especially protected areas remains lower than the average for the Ukraine. The list of negative parameters and detrimental influences on the environment could be continued.

However, at this stage it should be pointed out that human beings also suffer as a result. I would like to emphasise a single parameter – the mortality rate among children and teenagers in the region has increased by 38 percent over the last six years.

Since 1997, the obligatory regulations on the "Protection of natural environment and provision of ecological safety in the region" have been included in the structure of regional Programmes of Economic and Social Development in the Donetsk Region. Without implementation of the measures laid down in these regulations the environmental parameters of the region would be much worse. Unfortunately, it is not possible to spend more than 50-60 million hrivnas a year on dealing with the environmental problems of the region, even if all available financial sources are used.

At the moment, a draft of the ecological target programme of the region is being developed and should be completed in 2005. During this period it will be necessary to solve even more complicated problems regarding the provision of ecological safety as, simultaneously with the growth of manufacture since 1999, emissions to the atmosphere, waste products and other negative influences on the environment have also increased. The target of the programme is not only to constrain the growth of negative influences on the environment, caused by the growth of manufacturing, but also to achieve their reduction. The achievement of this purpose becomes the basis of sustainable ecological development in the Donetsk Region.

Realization of the regional ecological Programme '2005 within the stipulated financial limits will allow us to reduce the emissions of hazardous substances to the atmosphere from 1.7 million tons in 2000 to 1.35 million tons in 2005. However, this can be achieved only if environmental norms for the construction of new plants and decommissioning of old manufacturing industries are strictly observed. The implementation of the Programme will also permit us to extend the area covered by the region's nature reserves to 4.1 percent in 2005.

The achievement of all the expected parameters with the 514 measures of the ecological Programme '2005 requires 1.2 billion hrivnas according to the levels of the technical decisions of 2000. Certainly, an increase in annual investment in measures of the ecological programme will be essential. However, financial opportunities for the realization of programmes are often limited. Therefore, the scientific and technical search and realization of economic decisions on environmental problems are the constructive mechanisms of achieving the target set by the ecological programme to be completed by 2005.

The search for economic decisions should be developed in two basic directions – a technological one, connected with search and realization of resources and energy saving and low wasting technologies that will automatically lead to a decrease in the technogenic load on the environment. The developers of such technologies should allow for ecological efficiency as a positive asset of these technologies when they are first introduced.

The second direction is the scientific and technical search for economic decisions in favour of nature protection devices and the technologies which prevent and neutralise hazardous emissions to the atmosphere, dumps to reservoirs, industrial and municipal wastes from existing and planned industrial and municipal facilities.

The programme "Scientific and technical development in the sphere of protection of the natural environment, ecological safety and reproduction of natural resources of the Donetsk Region" should become an integral part of the programme of scientific and technical development of the Donetsk Region planned to be completed by 2020.

The programme should provide a detailed approach to questions regarding the protection of the environment in Donbass. The major scientific and technical directives and tasks aimed at protection and augmentation of natural resources and provision of ecological safety in Donbass should be formulated on the basis of a scientific analysis of the state of the environment which includes parameters for the state of public health and a forecast of changes which may occur in the technogenic impact on the environment during a structural reorganisation of the regional economy.

It is possible to define the following environmental targets among the major objectives for the scientific and technical development of the region:

- development and implementation of a system of monitoring the environment;
- creation of a regional ecological information and analytical system;
- development of an informational and analytical database on dangerous waste products;
- creation of the centre of natural resources and investigation of pollution sources;
- development and implementation of ecologically safe production and protection equipment;
- search and implementation of low wasting and resource-saving technologies;
- detailed estimation of changes in the ecological situation caused by the negative consequences of the closure of collieries;
- involvement of the population in dealing with environmental problems as demanded by the Arhus Convention;
- search for new organisational and economic mechanisms for providing ecological safety in the region.

The Donetsk region has one of the greatest scientific potentials in the Ukraine. Six institutes of the National Academy of Sciences of Ukraine, 17 institutes of higher education, more than 100 research, design and technical institutes operate here. More than 26 thousand persons, amounting to 1.4 percent of the total number of employees in the region, are employed in the field of "science and scientific provision".

The scientific institutes deal with the problems of environmental protection, carrying out the appropriate scientific research, developing and designing documentation for the objects that present an ecological threat to the population and to nature. Appropriate informational and analytical systems purporting to technology, equipment and means of control of environmental protection and ecological safety as they relate to this specific region, are gradually being created. The projects and schemes of optimum environmental management are being developed. Research on preservation and reproduction of biodiversity is also carried out in the region.

I would like to give a brief account of some environmental problems which require scientific and technical investigation.

Atmospheric air protection

The problem of prevention of sulphur oxides emissions during the burning and coking of coals on an economical basis is as yet practically unsolved on an industrial scale. For Donbass with its huge amounts of burning and coking of coals containing up to 4 percents of sulphur, this is a major problem both in terms of volume and toxicity of such emissions.

22

Work on improving this situation has already begun. For example, work has started at the Starobeshevky Thermal Energy Station on the reconstruction of power unit No. 4, replacing the existing boiler TP-100 with a boiler burning coal and industrial wastes in a circulating boiling layer, thus reducing emissions of sulphur and nitric oxides to the atmosphere.

At Avdeevsky Coke and Chemical Plant the reconstruction of a plant of sulphur purification with deep purifying of coke gas from hydrogen sulphide (reduction from 10 to 0,5 g/m^3) is being carried out on the basis of Danish "Khaldor Topse" devices for purification of outlet gases and production of sulphuric acid.

All-round support of this work, including scientific and technical assistance, is necessary for the subsequent optimization and duplication of results in the Ukraine.

Conclusions: protection of water resources and handling of dangerous wastes

The problem of how to prevent dumping of highly mineralized mine waters into surface reservoirs on an economical basis remains unsolved. The first steps in this direction were made by "Stirol" Concern where the device of return osmosis for demineralization of sewage with a productivity of 1,000 m^3 an hour has been installed.

A search is being conducted for effective and economic ways to neutralise and recycle highly toxic wastes from chemical, coke and other plants, such as the toxic waste from the Gorlovsky chemical plant or stores of liquid toxic waste products from coke factories, accumulated and drained into underground horizons. The common Ukrainian problem of neutralization of accumulated forbidden and poisonous agricultural chemicals is not solved in the region. The common problem of ecologically safe processing of municipal waste should also be addressed.

At the same time, it is necessary to note that the operation of devices for processing liquid toxic wastes in Mushketovsky refuse dumps from Yasinovsky coke and chemical factory with production of a merchantable commodity output has begun. The site for processing of 150 tons of slam wastes of galvanic shop of station of neutralization is constructed at Torez electric and technical factory.

There are many similar environmental problems in most areas of industrial and municipal activity in the region. They are numerous in the field of restoration and reproduction of degraded natural resources. And those islands of a protogenic nature which have remained in the Donetsk Region still require our care and protection.

The development and implementation of the ecological chapter of the programme of scientific and technical development of the region by 2020 making use of the scientific and technical potential of the Donetsk Region in interaction with the highest potential of the National Academy of Sciences of the Ukraine will promote the solving of the environmental problems which have been accumulating for many years in the Donetsk Region.

An extremely important initiative in this field is the initiative of the Presidium of the Ukrainian Academy of Sciences and the Donetsk Regional Administration which has created the scientific and technical site "Ecology of Donbass", the prime task of which is the realization of principles of ecologically sustainable development in Donbass.

CHAPTER 3

The Impact of Metallurgical-Mining Enterprises on the Transformation of the Biosphere and Noobiosphere

Prof. Igor Khodakovsky, Valeria Strokova, Elena Zhuravleva

Chemistry Department
"Dubna" International University for Nature, Society and Man
Universitetskaya Str. 19, Dubna
Moscow Region 141980
Russia

Summary

This chapter examines the impact of metallurtigan and mining enterprises in the biosphere and noobiosphere with a special emphasis to their transformation.

Introduction

In the second half of the 20th century, due to the development of industrial society, the scale and intensity of man's pressure on ecosystems has essentially changed. The biosphere's new state named 'noobiosphere' by I.L.Khodakovsky in 1995, has appeared [Khodakovsky I.L. The Noosphere conception..., Khodakovsky I.L. Noosphere – the modern...]. In this new term man's intellectual role is underlined as the determining factor of nature transformation, and at the same time a close link with the traditional term "biosphere" is maintained. Thus, noobiosphere is examined as a transitive state of biosphere to the state, which is either closer to the ideal one, or, in contrast, is farther away. It may be possible to speak about a "noo" biosphere, formed in the second half of the twentieth century, in the sense that the inverted commas in general reflect the unreasonableness of modern humanity's collective intellect. In spite of this, the term "noosphere", widely used by V.I.Vernadsky and by his followers should be understood as the biosphere state which is characterized by the ideal interrelations of a human being and nature. It is obvious that this sort of state cannot be achieved in essence, though this concept is rather useful.

Towards the beginning of the third millennium the extent of human activity started to become comparable with geological processes. At local and regional levels the "technical pressure on nature becomes comparable with the natural processes (see table 1) and sometimes begins to overcome the buffer volume of the natural migration cycles, creating the integral effects leading to irreversible ecological consequences" [Naumov G.B. "The noosphere...].

Every year in the world people discover and process about 4.5 billions tons of various mineral resources. Less than 10% of them are actually used, and the rest is wasted. Useful ways of utilizing them are not usually determined [Strokova V.V. Evolution of Raw...].

W. Leal Filho and I. Butorina (eds.),
Approaches to Handling Environmental Problems in the Mining and Metallurgical Regions, 23–26.
© 2003 *Kluwer Academic Publishers. Printed in the Netherlands.*

Table 1- Local natural and man-made displacements of matter

The Kuril Islands 1200 km^2	Krivoi Rog 75 km^2
32 volcanoes (from 1930 till 1963	7 open pits, 8 underground mines
2.6 km^3 of material erupted.	(from 1953 to 1961 not less than
Average annual – 0.06 km^3/a year)	2.2 km^3 of rock extracted.
	Average annual – 0.08 km^3/a year)

This is why now, as never before, there is a danger of vast amounts of man-caused products accumulating. About the same amount of raw materials is mined every year for the needs of the building materials industry. The problem of wastes forming man-made landscapes causes, in its turn, the problem of large-scale ecological imbalance.

Human technogenic activity has profoundly altered the flows of chemical elements in the biosphere. As the result of this, the chemical composition of the atmosphere, hydrosphere and topsoil of the planet has changed drastically. In the early 20th century V.I.Vernadsky wrote, "The last centuries have revealed the new factor which increases the number of free chemical elements, mainly gases and metals on the earth surface. This factor is human activity" [Vernadsky V.I. Izbrannye sochineniya…, p.2., p.35].

Hydroelectric development and the formation of large water basins on the big rivers, the position of large metallurgical-mining plants on river banks and on the coasts of the lakes and seas (Azov, Caspian, Black and others) coupled with other large-scale technogenic forces have resulted in changes in bio-geochemical cycles of biogenic macro- and microelements and, thus, have diminished the stability of ecosystems in the biosphere. The global sulphur and nitrogen streams, caused by human activity, have now become comparable in size to natural streams of these elements and even exceed them [Clark W.C. Managing Planet Earth…]. With regard to metal present in trace concentrations, many of these are toxic for every living thing, according to J.Nriagu and J.Pasin [Nriagu J.O. and Pacyna J.M. Quantitative…], industrial wastes of lead, cadmium and zinc exceed the natural sources of these metals by 18, 5, 3 times respectively. As for other metals, including arsenic, mercury, nickel and vanadium, the volume produced by human activity in the global streams nowadays is double that found in natural sources.

The aspiration towards military and political superiority, the excessive centralization of authority and financial flows, and also "negative" selection of people in managerial positions as well as the "police" censorship of mass media in the Soviet Union have resulted in catastrophic erroneous and incompetent administrative decisions at all levels. In the last decade during the rapid political and economic changes in Russia, a great number of decisions have been accepted by the legislative and executive authorities. However, the conformance of the decisions to criteria of reasonableness, optimality and practicability are often determined by power interests, because the society as a whole is possessed by a passion to make profit "at any price". The degradation of society and nature cannot be stopped without the creation of a civil society which understands clearly the essence of the ecological problems. The latter, in turn, is not possible without both the revision of the educational system and the development of high technologies. Thus, it is important to emphasise that moral self-restriction is necessary in employing scientific and technological achievements on the path towards a perfect system biosphere-human society.

As G.B.Naumov stressed, "the problem of industrial wastes conversion into the new natural resources (to transmit the technogenic rubbish dumps into technogenic deposits) cannot be solved either by the ecologists alone, or by the technologists alone" [Naumov G.B. "The noosphere..."]. The task of the secondary use and industrial wastes buring is not solved without counting with migration laws, dispersion and concentration of elements in noosphere. The forms of elements being in the environment determine their influence upon the living matter, including upon humans.

The major problem arising when looking for a solution to environmental problems, connecting complex ecomonitoring and management of water resources, is the consideration of the discrepancy between the natural borders of ecological regions and historically formed administrative borders of the regions involved. The latter could be distinguished within the frame of separate country (district, area), as well as within the frames of the political unions in adjoining states. Ecological region (ecoregion) is defined by V.A.Kovda as the catchment basin in its hydrographic borders (including underground waters). The largest river basins can be divided into subbasins. The advantage of a catchment basin as against other space units is that it has uniform directions of substance flows (including pollution) to the base level of erosion. The latter facilitates the computer modelling of biogeochemical cycles of chemical elements including the mass transfer in space and time coordinates.

The ecological safety of ecoregions with large metallurgical-mining plants is clearly only possible if complex prognoses can be provided. These should consist of databases of various contents, and should be functionally connected by information channels both with the modelling systems and with the geographic information systems (GIS). The development of these prognoses should be based on the treatment of multiple-factor and dynamic biosphere processes that change constantly under the influence of human activity. Imitating computer modelling (numerical experiments) enables us to predict tendencies towards changes in ecological characteristics in ecoregions with large metallurgical-mining enterprises. Thus, it becomes possible to predict the after-effects of actions aimed at the environmental restoration of contaminated regions.

Modern databanks containing ecological information are based on computer geographic information systems. GIS integrate diverse data (geographical, hydro-geological, geochemical, medical, demographic, social, economic and others) and data concerning the technogenic pollution of regions. These integrated systems should give both objective status information on the environment at global, regional and local levels and objective predictions of any long-term ecological variations. Computer maps created by such systems in the near future should become obligatory documents for authorities making administrative decisions concerning water use and social protection of citizens.

Computer ecological information systems for administrative regions (and ecoregions) should be available on three interconnected levels each of which is determined by the problems which have to be solved and the qualifications of the users:

- Systems with detailed information for research work. These should be on line with custom GIS information and should include the databases containing the physical, chemical and physicochemical constants needed for mathematical modelling of noobiosphere contamination by toxic substances in the ecoregions with large metallurgical- mining plants;

- Expert information systems developed for authorities, engineers and other users who make the relevant decisions on the use of natural resources;

- Systems giving authentic and objective information for representatives of authorities at local, regional and federal levels, and also for citizens, educational institutions and public ecological organisations.

Obviously, the degree of aggregation of the information should increase from the first to the third level of the information systems mentioned above. The majority of the information and computing opportunities of the first level systems should be presented in the second, whereas on the third level it is expedient to present demonstration versions only. However, despite the restrictions of the second and the third level, the conclusions presented there, although not containing numerous details, should be clear enough for experts to make the necessary decisions, while non-professionals would find answers to a wide range of ecological questions.

References

Clark W.C. (1989) Managing Planet Earth, in *Scientific American v.261, N3, 47-54, 1989.*

Khodakovsky I.L.(1995) The Noosphere conception and problems of noogeochemistry. In: *Basic directions of geochemistry* - centenary of Academician A.P.Vinogradov - Moscow, "Nauka", ed. E.M.Galimov, pp. 289-299.

Khodakovsky I.L. (2001) Noosphere – the modern state of biosphere. // Interstate conference "The scientific inheritage of V.I.Vernadsky in the context of civilization global problems", 23-25 of May 2001, Reports. – Moscow.: Enterprise Press, "Noosphere", 2001. pp. 51-66.

Naumov G.B. (2001) The noosphere studies development. // Interstate conference "The scientific inheritage of V.I.Vernadsky in the context of civilization global problems", 23-25 of May 2001, Reports. – Moscow.: Enterprise Press, "Noosphere", 2001. pp. 15-28.

Nriagu J.O. and Pacyna J.M. (1988) Quantitative assessment of worldwide contamination of air, water and soils by trace metals, Nature. 1988. V.333, pp.134-139.

Strokova V.V. (2000) Evolution of Raw Materials Evaluation Criteria in the Industry of Building Materials. Contemporary problems of science of building materials: 4th Academic hearings/ Ivanovo State architectural-building academy. - Ivanovo, 2000, pp. 511-514.

Vernadsky V.I. Izbrannye sochineniya, v.1-5, M. Izd-vo USSR Academy of sciences, 1954-1960.

CHAPTER 4
Impact of the metallurgical industry on the coastal ecosystem of Black Sea countries

Dr. Irina Rudneva

Institute of the Biology of the Southern Seas
National Ukrainian Academy of Sciences,
Nahimov av., 2, Sevastopol 99011
Ukraine

Summary

Over the past 40 years the Black Sea ecosystem has been in a state of environmental crisis resulting from the high anthropogenic impact. The effects of contamination by, amongst other things, heavy metals, oil, detergents and pesticides have lead to negative ecological consequences. Pollution from riparian state operations, discharges and dumping from port zones are together contributing to the degradation of the Black Sea coastal shelf. The main sources of heavy metals are industrial effluents from the ferrous and non-ferrous metallurgical industry. High concentrations of Cu, Pb, Zn, Cr, As are present in the sewage of the metallurgic industry. All these pollutants are distributed in the marine water, accumulated in the lower sediments and marine organisms. The presence of heavy metals in the water and sediments change their composition and properties resulting in unfavourable conditions for biota. Furthermore, heavy metals accumulate in the marine organisms leading to their death, deterioration of health, and loss of biodiversity.

Introduction

It is estimated that 162 million people live in the Black Sea drainage basin. Almost one third of the entire land area of continental Europe drains into the sea. Besides this, the drainage basin of the Black Sea includes major parts of 17 countries, 13 capital cities and a lot of industrial and port centres. The second, third and fourth major rivers such as the Danube, Dnieper, Don, Dnister discharge into the Black Sea. Sewage from the industrial, agricultural, transport and domestic sectors is the main contributor to the pollution of the Black Sea. Additionally, there is a high concentration of mining industry operations in the Black Sea basin. It is common knowledge that the mining industry is responsible for large amounts of wastes. For example, in the process of production of 1 ton of metal such as Au, Ag or Pu, 5-10 tons of wastes are produced during the mining cycle and 10-10 000 tons during processing (Mining and environmental, 2001). Thus, the main source of chemical pollution (heavy metals, phenols, oil), is, undoubtedly, the effluents from metallurgical, mining and additional industrial activities (Mee, 1992; Aybak, 2001; Rudneva, Petzold-Bradley, 2001). At the present time the Black Sea is suffering from a catastrophic degradation of its natural resources as a result of the massive anthropogenic impact which it has experienced and which includes industrial activities in riparian and non-riparian countries.

W. Leal Filho and I. Butorina (eds.),
Approaches to Handling Environmental Problems in the Mining and Metallurgical Regions, 27–33.
© 2003 *Kluwer Academic Publishers. Printed in the Netherlands.*

Effluents as sources of pollution in the Black Sea and Sea of Azov

There are 14 branches of ferrous and 10 branches of non-ferrous industries in the Black Sea Basin. The sewage from these industrial activities is an important source of heavy metals and organic substances (Table 1).

**Table 1. Characteristics of the metallurgical industry in the Black Sea drainage basin
(Fashchuk & Sapozhnikov, 1999)**

Industry	Number of branches	Production	Pollutants
Ferrous	14	Non-carbon ferrochromium, Hard ware, coking chemistry	Cr^{6+}, oil
Non-ferrous	10	Pb, Ti, Cu (anodic, catodic)	phenols, Pb, Zn, As, Mn, Cu

The effluents of ferrous and non-ferrous industries contain a lot of persistent pollutants which are contributing to the degradation of the coastal marine ecosystems. The most dangerous biological toxicants are present in high concentrations in the metallurgical industry effluents: Cu (200-350 mg/l), Pb (2860 mg/l), Zn (220-850 mg/l), As (140-1000 mg/l), phenols (200 mg/l) (Fashchuk. & Sapozhnikov, 1999). In addition to this, the Danube alone discharges 1000 tons of Cr, 4500 tons of Pb, 6000 tons of Zn, 900 tons of Cu, 60 tons of Hg and 50 000 tons of oil annually (Mee, 1992).

According to recent data (Fashchuk & Sapozhnikov, 1999) the main source of copper, lead, chromium and zinc is sewage from industrial plants in Odessa, Sevastopol, Kerch, Mariupol, Novorossiysk, Trabzon, Samsun, Zonguldak, Varna, Konstatza, Bratislava, Tiraspol, Mogilev, Kremenchug, Dneprodzerjinsk, Burgas, Budapest, Tiraspol, Rostov-on the Don, Krasnodar and Zaporohie. The main source of phenols is the sewage from the metallurgical industry in Odessa, Istanbul and Linz. Besides this, combinative pollution is present in the effluents of ferrous and non-ferrous industries. For example, the Sevastopol plant " Dominanta" which processes scrap metal contaminates the Inkerman Bay and the marine environment. In marine water in this area the concentration of Fe is more than 1.5-2 –fold higher and the content of oil is more than 1.2-6 – fold higher than in the other bays of the Sevastopol coastal area (according to the data of the State Inspection of Black Sea Security, Sevastopol branch).

There are 16 official dumpings in the western part of the Black Sea. Approximately 10% of pollutants are the products of 11 dumpings. Most of these are situated on the continental shelf, and they are used for the disposal of dredging spoils (sludge). Sludge from different land sites contains toxicants from industrial effluents and dumping results in these materials entering sea water (Mee, 1992). The mining plant in Balaklava Bay (Sevastopol) drains daily 11 500 m^3 effluents into the shelf zone. This has catastrophic consequences for the coastal ecosystem and in particular for biota (Kuftarkova et al., 1999). Sevastopol Bay is also highly polluted because 9 tons of heavy metals, 19 tons of oil and 96 tons of detergents enter it annually through industrial and port effluents (Shadrin, 1999).

Accidents in mining and metallurgic plants have catastrophic consequences for ecosystems, biota and human health. On 31.01.2000 the accident in the gold mining plant in Romania resulted in 100 000 m^3 effluents containing cyanides being drained into the river Samosh and from there into the Black Sea. The concentration of toxicants in the effluents was more than 800 times higher than the maximum legal levels. Such pollutants are distributed in river and sea waters and accumulate in the bottom sediments. This causes negative biological effects on river and sea ecosystems. The number of fish and other aquatic organisms declines, and human needs can no longer be met.

Heavy Metals Pollution the Black Sea and Azove Sea

Almost all large cities in the Black Sea basin and, especially port zones, have all kinds of metallurgical and mining plants. Industrial effluents enter the Black Sea through the rivers. Thus, high concentrations of Pb, Cu, Zn, Mn, As and Cr are detected in Black Sea and Sea of Azov water and sediments (Hydrometeorology and hydrochemistry.., 1996). In the bottom waters of the northwestern part of the Black Sea they are estimated at 0.10 gµ/l Hg and 1.0 gµ/l Cr. The highest concentration of As (2.29-17.13 µg/l) was identified in the Sea of Azov. This data is higher than the maximum legal levels (10.0 µg/l). As for Hg and Cr contents they are evaluated as 1.0-1.0 and 1.0-3.7 of maximum legal levels. The highest concentration of heavy metals accumulated in sediments was detected in the region of Mariupol (Sebah, Pankratova, 1999).

Mercury is widely distributed in water and sediments, and its concentration in sea upper layers is more than 1 µg per l. The total Hg content in Black Sea upper layers is probably approximately 263 t (Applied ecology..., 1990). Thus, high concentrations of heavy metals in water and sediments mean that they accumulate in seafood, which is very dangerous for human health. High pollution of heavy metals was determined in the shelf zone, where their content in bottom sediments was distributed as follows, Hg<Co<Pb<Cu<Zn<Cr<Sr; and in marine water, Hg<Co<Cr<Pb<Cu<Zn (Eletzkiy & Hosroev, 1992). At the same time pollutants migrate all over the sea, they are transferred to and become concentrated in the sediments of otherwise non-polluted areas. For example, concentration of heavy metals in Jalita Bay (Crimea) in 1992 was estimated as follows, (mg/l) Cu –1.2-2.4; Cd-0.1-1.0; Cr – 0.01-0.4; Fe –3.0-43.0; Zn –77.0-500.0 although there is no industrial activity in this region (Hydrometeorology and hydrochemistry.., 1996).

In industrial regions and port zones the content of heavy metals and organic compounds is higher. Concentrations of heavy metals in the water in Sevastopol Bay are estimated to be as follows: (µg/l): Hg – 0.01-0.74; Cr, 1.6; Fe 10.0; Ni –10.2;, Cd- 0.06; Cu –1.7-2.1; Zn –0.8-5.2(Shadrin, 1999). Thus, high concentrations of heavy metals in water and in sediments alter their properties and have a negative influence on biota.

Heavy Metals Accumulation in Marine Fish and Their Effects

Heavy metals accumulate in marine organisms' tissues and have negative biological effects on fish health and biodiversity (Table 2).

According to the data presented, the concentrations of heavy metals in some fish samples are higher than the maximum legal levels. This has a negative effect on fish health, because the high content level of these toxicants damages organs and tissues and stimulates inflammation, resulting in death, loss of biodiversity and stock of fish and other marine organisms (Table 3). Heavy metals and organic compounds transfer through food chains, and their toxic effects strengthen from one chain to another.

Table 2. Heavy metals content in some marketable fish tissues in the Azov and the Black Sea, mg/kg wetweight (Sebah et al., 1995)

Metals	Maximum legal levels	Fish species				
		Russian Sturgeon	Star Sturgeon	Mullet	Flounder	Perch
Cu	5.0	17.7	0.05-10.5	0.17-34.4	3.36	0.13- 6.4
Zn	40.0	71.7-97.0	26.9 -70.7	0.7-35.8	60.0	0.32-51.2
Hg	0.4		0.08	0.02-0.07	0.21	0.01-0.17
As	5.0		0.93	0.01-0.44	2.29	0.01-0.09
Pb	1.0		0.39	0.01-0.52	0.46	0.03-0.44
Cd	0.2		0.05	0.001-0.12	0.18	0.01-0.03

Table 3. Biological effects of metals on fish

Metal	Biological effects	Cited literature
Ni	Cytotoxic, gonadotoxic, carcinogenic, embryotoxic effects. Damage to eyes, inflammation, tissue necrosis	Acevedo et al., 2001; Ermolli et al.,2001; Moiseenko, 1999
Cu	Damage to organs and tissues, anaemia, changes in the processes of blood metabolism, hemopoesis, gills, respiration and behaviour	Sebah et al., 1999 Dethloff et al., 2001
Al	Damage to metabolic processes in tissues and organs, cell necrosis	Moiseenko, 1999
Sr	Pathology of bones	Moiseenko, 1999
Pb	Damage to erythropoesis, damage to sensitivity and locomotion	Sebah et al., 1999
Zn	Degradation of gill epytelium, kidneys pathology, changes in behaviour, anomalies of growth and development	Sebah et al., 1999
Cr	Cytotoxic effects	Ermolli et al., 2001
Co	Cytotoxic effects	- " -
As	Damage to tissues	Sebah et al., 1999

Furthermore, mixing of toxicants has a negative influence on molecular defence systems of marine organisms and leads to changes in their health and biodiversity (Oven et al., 1998; Rudneva, 1998 a,b). This causes development anomalies, unsuccessful reproduction, modifies population structure and worsens health. According to the present studies, 143 fish species inhabit the Black Sea, most of which are to be found in the coastal area. During the period from the 1950s to the 1990s ichthyophauna was modified and fish biodiversity declined: in Sevastopol bays the number of fish species was reduced more than 2-fold and fish stock declined more than 100-fold as compared with the 50s (Konovalov, 1996; Oven,1993).

Furthermore, heavy metals which accumulate in seafood are a great risk to human health (Table 4). They modify metabolism, stimulate different pathologies, have a negative influence on children's growth and development. High concentrations of heavy metals and organic substances in the air provoke pathologies of the respiratory organs. According to the official statistics for 1997 in Inkerman (the industrial region of Sevastopol, where more than 20 industrial plants are situated, including metallurgical and mining industries) pathology of the respiratory organs in children (for 100 000 people) was detected in 64,424 cases as compared with 50,745 in

Sevastopol, pneumonia in 1345 and 407 cases, respectively. Chronic bronchitis in adults was estimated at 168 cases in Inkeram and at 67 in the city; in children the estimated figures were 747 and 139 respectively.

Thus, the Ukraine loses approximately 300 million dollars annually as a result of environmental pollution, unsustainable development, non-optimising   application of natural resources, and inadequately considered decisions. As the present information shows, it is essential to improve and expand on the basis of coordination and co-operation in order to prevent future pollution of the Black Sea by heavy metals and organic compounds contained in the effluents from the metallurgical and mining industries. In essence the concept of sustainable development should be applied to adaptive management, better linkage between monitoring systems, improved knowledge of pollution sources and pollutants in the Black Sea and Sea of Azov basins, their effects on biota and risk assessment for ecosystems and human health. Political, economic, social and scientific collaboration, Internettechnology and ecological education could all be developed as important mechanisms for the restoration of the ecosystem.

Table 4. Biological effects of metals on human health

Metals	Biological effects	Cited literature
Hg	Neurobiological effects, damage to brain and nervous system	Moiseenko, 1999
Cd	Gonadotoxic, carcinogenic effects	Moiseenko, 1999; Lenihan, Fletcher, 1976
Pb	Damage to mental performance and Development, growth and hearing, reduction of haemoglobin synthesis, kidney pathology	Hens et al., 1998
Ni, Co, Cr	Allergic dermatitis, irritant reactions on the skin, cytotoxic effects	Ermolli et al., 2001
Sr	Bone pathology	Moiseenko, 1999
Cu	Anaemia, reduction of haemoglobin synthesis	Moiseenko, 1999

References

Acevedo F., Serra M-A., Ermolli M., Clerici L., Vesterberg O. (2001) Nickel-induced proteins in human HaCaT keratinocytes: annexin II and phosphoglycerate kinase. Toxicology, 2001. Vol. 159, Is. 1-2, pp.33-41.

Aybak, T. (2001)Globalization in Europe and new regionalism in the Black Sea: Towards innovative policy in environment. *Perspectives of integration and development of R & D and the innovation potential of the Black Sea Economic Cooperation Countries (BSEC).* NATO Advanced Training Course, Yalta, Ukraine, Oct. 29-31, 2001, pp 46-58.

Dethloff G.M., Bailey H.C., and Maier K.J. (2001) Effects of dissolved copper on select haematological, biochemical and immunological parameters of Wild Rainbow Trout (Oncorhynchus mtkiss) Arch. Environ. Contam. Toxicol. 2001. V.40. , pp. 371-380.

Eletskiy, B.D. and Khostroev, V.V. (1992). Anthropogenic pollution in the inshore areas of the Black Sea in summer 1989. *Ecology of the Black Sea coastal areas: Collected papers.* Moscow: VNIRO. 234-250 (In Russian).

Ermolli M., Menne Ch., Pozzi G., Serra M-A., and Clerici L.A. (2001) Nickel, cobalt and chromium-induced cytotoxicity and intracellular accumulation in human hacat keratinocytes. Toxicology, 2001. V. 159. Is. 1-2. , pp. 23-31.

Fashchuk, D.Ya. and Sapozhnikov, V.V. (1999) Anthropogenic Load on the "Sea Watershed Basin" Geosystem and its Consequences for Fisheries (Methods of Diagnosis and Forecasting on the Black Sea Example). Moscow: VNIRO Publishing. (In Russian)

Hens L., Melnil L., and Boon E. (1998) Environment and human health. Kiev:Naukova Dumka. P. 303

Keonjan V.P., Kudyn A.M., Terehin U.V.(eds) (1990) Applied ecology of marine regions. Black Sea Ed.: Kiev: Naukova Dumka. (In Russian)

Simonov A.I., Raybinin A.I (eds) (1996) Hydrometeorology and hydrochemistry of seas. V.4. Black Sea. Iss. 3. Present situation of Black Sea pollution. Sevastopol:ECOSY-Hydrophysics. (In Russian)

Oven L.S. (ed) (1993) Ichthyophauna of Black Sea bays under the anthropogenic impact. Kiev: Naukova Dumka. (In Russian).

Kuftarhova E.A., Kovrigina N.P., and Rodionova N.U. (1999) Hydrochemical characteristics of water in Balaklava Bay and its coastal area of Black Sea. Hydrobiological J. 1999. V.35, N 3., pp. 88-99.(In Russian)

Konovalov S.M (ed) (1996) Modern state of Black Sea ichthyophauna. Sevastopol: Institute of the Biology of the Southern Seas (in Russian)

Lenihan J. and Fletcher W.W. (1976) Health and the environment. Glasgow and London: Blackie 231 P.

Mee, L.D.(1992). The Black Sea in Crisis: A need for concerted international action. *AMBIO 21,* pp. 278-286.

Mining and environment. Moscow: Logos, 2001. (In Russian).

Moiseenko T.I. (1999) Evaluation of ecological risk in the cases of metal pollution in water. Water resources. 1999. V. 26, N 2., pp. 186-197 (In Russian).

Oven, L.S., Shevchenko, N.F., Rudneva, I.I. (1998) Response of Black Sea scorpion fish *Scorpaena porcus* on pollution in the marine environment. Proceedings of International Symposium of marine pollution. Monaco, Oct. 5-9, 1998. Extended Synopsis. 587-588.

Rudneva, I.I. (1998a) The biochemical effects of toxicants in developing eggs and larvae of Black Sea fish species. Marine Environmental Research. 46, 499-500.

Rudneva, I.I. (1998b) The responses of blood and liver antioxidant system in two Black Sea fish species as biomarkers of pollution effect. Proceedings of International Symposium of marine pollution. Monaco, Oct. 5-9, 1998b. Extended Synopsis. 618-619.

Rudneva I., Petzold-Bradley E. Environment and security challenges in the

Black Sea region. Environmental conflicts: Implication for Theory and Practice. Netherlands: Kluwer Academic Publishers. 2001. P. 189-202.

Sebah L.K., Pankratova T.M., and Avdeeva T.M. (1995) Evaluation of heavy metals accumulation in marine organisms in Azov-Black Sea basin. Proc. Southern Scientific Research Inst. of Marine Fisheries and Oceanography 1995. V. 41, pp. 87-90 (In Russian).

Sebah L.K. and Pankratova T.M.. Evaluation of Black Sea and Aqzov Sea pollution in modern anthropogenic status. Proc. Southern Scientific Research Inst. of Marine Fisheries and Oceanography 1995. V. 41, pp.91-93 (In Russian).

Shadrin, N.V. (1999) Function of ecosystem and economics: relationships in global and local levels. *Sevastopol aquatory and coast: ecosystem processes and services for human society* . Sevastopol: Aquavita, 1999. 17-24 (In Russian).

CHAPTER 5
Sustainable water management, a matter of looking forward

Dr. Peter J.H. van den Berg

Alkmaar University of Professional Education
Bergerweg 200
1800 AK Alkmaar
The Netherlands

Summary

Sustainable water management is part of the broader concept of sustainable development. This chapter presents an example of sustainable water management in the Netherlands and Belgium[1]. It emphasises the importance of definitions, concepts, principles and systems on the one hand, and of their contextual practical implementation on the other hand. Essential to the success of sustainable water management are integration, participation and co-operation[1].

Concept

In his *concept* of sustainable development Van Weenen formulates: 'Sustainable Development is an integral concept for achieving quality of life. Life is about "L" for Limits, "I" for interdependence, "F" for fundamentals and "E" for equity. This set of issues reflects the importance of dealing with material concerns, acknowledging the relationship between humanity and nature, being committed to addressing fundamental causes, and considering ethical values[2]. Humanity, therefore, must formulate totally new principles for fundamentally different systems of production and consumption.

Sustainable water management is defined as resource, context and a future-oriented production of water, aimed at the fulfilment of elementary needs, better quality of life, equity and environmental harmony.[3] Sustainable water management must be shared, guaranteeing basic needs for all, and it must be sustainable, without mortgaging the choices of future generations.

Context

Sustainable water management implies the whole domain of knowledge, concepts, theories, techniques and methods, procedures, organisations and processes which serve to prevent, resolve or reduce water problems. In this case the Dutch prince Willem Alexander launches the idea of a Water Management Information System[4]. The objective of sustainable water management is to develop, design, apply, evaluate and improve solutions that are aimed at meeting elementary needs on the basis of natural resources in their local and regional context. The basis of and central to sustainable water management are definitions, concepts, principles and systems on the one hand, and contextual practical implementation on the other hand[5].

Co-operation

Integration, participation and co-operation are essential to the success of sustainable water management. Van Weenen: This can be realized regionally by contact networking of proponents and pioneers, and internationally by virtual networking. Through integration of local and regional

W. Leal Filho and I. Butorina (eds.),
Approaches to Handling Environmental Problems in the Mining and Metallurgical Regions, 35–38.
© 2003 *Kluwer Academic Publishers. Printed in the Netherlands.*

considerations with universal principles and global ideas, an enormous potential of new and sustainable options can be developed. This will be demonstrated by the inter-regional project 'Water management in the Central Benelux area'.

History

About one thousand years ago a part of Holland called 'De Krimpenerwaard' consisted mostly of water. It was known as 'Lacke and Isla'. It had an open sea connection and was colonised by people who started to dig ditches. The effect of these ditches was that water flowed into them, and dry land was the result. Because of this withdrawal of water the landscape changed into peat land. The ground settled and bedded down and natural drainage was no longer possible because the land became lower and lower. The result of this process of reclamation was that too much water flowed into the ditches and the water threatened the land. Dams were constructed, but the situation remained. In these early days of water regulation, water management was necessary from a natural point of view. Therefore, the people who lived in this area of Holland, developed a system with wind(water)mills to regulate the water level. The first mill was built in 1411. With those mills it was possible to pump the water to other areas of that particular part of the country. This early system of water management was in use for over 500 years and enabled the farmers to control the groundwater level.

A safe and habitable country with healthy and sustainable water systems is the aim of the Netherlands Directoral for Public Works and Water Management, established in 1798. After all, the nature of our low-lying, waterlogged country is such that it must constantly be protected against flooding from the sea and the rivers. At the same time, constant effort has to be made to secure the consolidation of the soft subsurface in order to keep the country habitable and cultivable.

The Inter-regional project 'Water management in the Central Benelux area'

Now in the 21st century in the Netherlands and Flanders we know that our supply of ground water and surface water is not endless, but limited[6]. In the Netherlands, for instance, water consumption has increased to 1.8 billion m^3 water a year. In general, industry consumed 510 million m^3, while private households used 700 million m^3. The rest was used by commercial, public and tourist facilities such as swimming pools and so on. In the Netherlands 25% is surface water, 75% is groundwater. Overall in Flanders 840 m^3 was used in 1997. 32 % was surface water while 56 % was groundwater. Although the Netherlands and Belgium belong to the most water rich countries in the world, problems are ahead. So water conservation is one of the main issues in these countries.

The aim of the 'inter-project water management in the Central Benelux area' is to prevent the drying out of land in the south of the Netherlands and Belgium. The project aims to handle water in the agricultural sector economically, i.e. to ensure that water is retained as long as possible.

Aim of the project

Because ground water is scarce each sector of society has to handle it sparingly. Tackling this problem in an integral way in the agricultural sector by conserving water, by water management and integration of the theme 'water' in management in general, will optimise the water balance and in so doing make it possible to reverse the reductions in the ground water level.

The management of water resources, which are so vital to life, must be tackled in an appropriate manner. Farmers and market gardeners play an important role in this process. After all, the important process of water conservation takes place on their land, and it is there that they can contribute to responsible water utilisation.

Nine Flemish and Dutch partners in four provinces have joined together in the project 'Water management in the central Benelux area'. This unique project began as the result of developing and exchanging knowledge. It is meant to help farmers and market gardeners in taking measures to maintain the water level at an appropriate level. All the project partners are examining possibilities of retaining water for a longer period of time on farmlands.

The project [7] 'Water management in the Central Benelux area' focuses on practical measures that can be applied by farmers and market gardeners. In addition, technical and scientific knowledge is being developed in the area of 'water conservation and water level management' and 'water management at business level'.

Water conservation and water level management

The aim of water conservation is to arrange test areas for water conservation in the south of the Netherlands and Belgium. In these areas measures have been taken to keep the water situation both with regard to ground and surface water at an appropriate level.

If water is retained for a longer period of time on farmland, it has more time to sink into the ground. This water is then added to the local water supply. As a result the water level rises. During the winter and before the growing season, the rainfall is not needed for the crops. By constructing dams, raising the level of the bedding ditches and filling the channels, it is possible to retain the water in the area for a longer period of time. During the growing season the water that has been retained can be used for irrigation of crops.

During the course of the project water conservation measures will be implemented in thirty test areas and the results will be closely monitored. We expect that the water conservation measures may have positive effects on the flora and fauna.

Water management at a business level

The amount of water used for optimising crop growth depends on a number of factors. One of these is the groundwater level. Other important factors are the type of crop grown, the type of ground, precipitation level, and the evaporation due to the sun.

At 80 farms and market gardening companies, the already existing methods for calculating the amount of water to be sprinkled on the crops will be further developed and refined. These methods help us to utilize water economically and sparingly.

Discussion

In this project we employed the Dutch water spray system which gives the farmers the choice of when to use it or not, and the Belgium water spray system which is highly centralised. Both systems had positive results in their respective agricultural situations and they have both contributed to a more economic way of handling water resources in the central Benelux area.

Notes

1. J.C. van Weenen, (2002) Concept, context and co-operation for sustainable technology
2. idem
3. J.C. van Weenen, (2000) Towards a vision of a sustainable university.
4. idem
5. Willem Alexander of The Netherlands, Global Water Partnership (GWP) at the International Conference on Freshwater, Bonn, Germany 2001. The Global Water Partnership (GWP), established in 1996, is an international network open to all organisations involved in water resources management: developed and developing country government institutions, agencies of United Nations. Its mission is to support countries in the sustainable management of their water resources.
6. J.C. van Weenen, (2002) 'Concept, context and co-operation for sustainable technology
7. Inhoudelijke eindrapportage Inrterreg-project Watermanagement in het Benelux Middengebied (B/98/TH3/05)

CHAPTER 6
Some facts on the links between human and environmental health in the Dniepro Region

Prof. Anatoliy Prykhodchenko, Valentina Karmazina, Paul Postnikov

Dnieprodzerzinsk State Technical University
Prospect Metallurgiv, 54/56
Dnieprodzerzhinsk 51940
Ukraine

Summary

This chapter presents the results of a study on populational health in the Dniepro region, Ukraine and discuss the implications of its findings

Introduction

It is widely known that the mining and metallurgical industrial complexes in both the Ukraine and elsewehere in the region are characterized by a high level of environmental degradation and that such degradation often influences people´s health. Yet, there are to date comparatively few studies which examine the links between environmental degradation from mining and from metallurgical activities, and human health. This matter is explored in this chapter.

A group of 150 persons from the Dniepro region were examined using a special haematological method (studying leucogrammes of venous blood leuconcentrate). The disturbances we discovered in the blood system of the population have the following indications of affection of granulocitary shoot: signs of destabilization in the system of immune competent cells; activization of the monocitary link of non-specific defence; venous blood plasmatisation; appearance of non-typical cell populations.

The concentration of non-typical cells in healthy people (1.88±0.86 %) differs greatly from that of other groups. Dnieprodzerzhinsk inhabitants (risk group) have the same level (8.46±5.02 %) of non-typical cells as those who took part in eliminating the consequences of the Chernobyl accident, their level of non-typical cellular elements being estimated the highest (11.59± 5.76%).

The data obtained give us every reason to think in terms of the so-called 'ecological AIDS' amongst the population living under high technogenic pressure.

Materials and methods of research

A group of 150 persons who live in the Dniepro region were examined. Four groups were formed to evaluate the health of the population with the help of haematological methods. The 1st control group consisted of 26 persons who could be considered healthy and had no complaints at the time of

W. Leal Filho and I. Butorina (eds.),
Approaches to Handling Environmental Problems in the Mining and Metallurgical Regions, 39–43
© 2003 *Kluwer Academic Publishers. Printed in the Netherlands.*

40

examination. The 2nd control group consisted of 31 persons who had been subjected to electromagnetic emission. 35 of those who had taken part in eliminating the consequences of the Chernobyl accident constituted the 3rd control group. The main group consisted of 31 persons, all of whom were Dnieprodzerzinsk inhabitants, and represented the people with occupational diseases (risk group).

A special method (leucogrammes of venous blood leuconcentrate - See Приходченко А.А., Исследование лейкоконцентрата...) was applied together with more conventional ones (See Меньшиков В.В. Руководство...). Statistical processing of the data obtained was carried out by application of standard methods.

Results and discussion

The venous blood leucoconcentrate (VBL) of those examined was found to contain 2-4 % of gigantic cell forms and segment nuclear elements in which granulation was disturbed. There were cases of both of absence of granulation and accumulation of toxic granularity. The system of mononuclear elements is involved in the pathological process, this system being able to respond to outer irritants by forming monoblast cells.

Pathology is accompanied by the involvement of the eritron system. We registered normoblasts, megaloblasts as well as rethiculomegaloblasts, though in small quantities (Fig. 1).

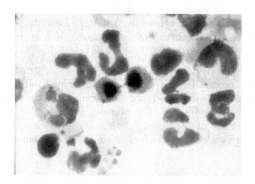

Fig. 1. Venous blood leucoconcentrate
Refract anaemia. In the centre – oxyphil normoblast in the state of division. Here and future: colouring made using the Romanovsky-Gimse method. Magnification 100 ·70

In the picture (Fig. 2) one can see a microscope photo of a bronchial lymphatic node smear. Coal particles contaminating the respiratory organs of the patient can also be seen.

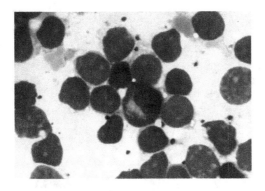

Fig. 2. The smear of a lymphatic bronchial node. Coal particles.

Immune system activation manifests itself as the tissue plasmatisation phenomena (Fig. 3). Cariorexis (Fig. 4) is one of the characteristic signs of strong influence on the cell genetic mechanism.

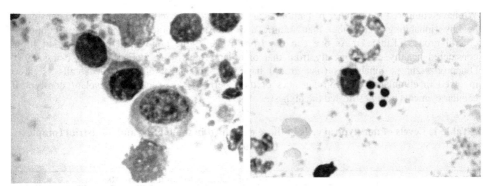

Fig. 3. The group of plasmocytes at different   *Fig. 4*. Cariorexis
                stages of differentiation

Non-differentiated blast cells can seldom be found in the VBL of people who can be considered healthy. When there are 5 or more percent of blasts in the VBL, complaints arise. In more serious cases where non-typical cells with deformed nuclei appear in the patient, they are considered hardly differentiated (Fig 5).

42

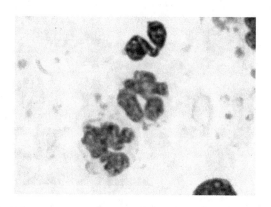

Fig. 5. Non-typical cells with deformed nuclei

The data obtained testify that the blood synthesis system damage mechanism renders effect at a rather high level of a stem polypotent blood-producing cell.

We have compared the number of pathological elements of the venous blood in those considered healthy (inhabiting areas other than mining and industrial regions) with the above-mentioned control groups (Table 1). As one can see from Table 1, the non-typical cells' level of those considered healthy differs greatly from that of other groups (the statistical level being 0.95). Dnieprodzerzhinsk inhabitants (risk group) have the same level of non-typical cells as those involved in eliminating the consequences of the Chernobyl accident, their level of non-typical cellular elements being estimated the highest.

Table 1. Levels of non-typical cells in those examined in the mining and industrial complex (risk groups)

Statistical terms	Results of examination of the residents			
	People considered healthy	Dnieprodzer-zhinsk residents	People subject to electromagnetic emission	Eliminators of consequences of Chernobyl accident
Average value	1.88	8.46	8.79	11.59
Average value	2.25	7.73	7.91	12.09
Dispersion	0.75	25.17	31.60	33.21
Standard deviation	0.86	5.02	5.62	5.76
Volume of sample	26	31	31	34
T revealed	-	6,592	6,195	8,494
F critical	-	2,004	2,004	2,002
T revealed	-	33,279	41,775	43,907
F critical	-	1,919	1,919	1,902

The data obtained give us every reason to consider the myelodysplastic syndrome (See Layton D.M. at al. The Myelodysplastic... p. 227-228) or so-called "ecological AIDS" in a population living under high technogenic pressure.

References

Приходченко А.А. Исследование лейкоконцентрата венозной крови в клинической экологии. Методика исследовательская - М 123.38.005-92. НПО «Вектор», 1992.- 28 с.

Меньшиков В.В. (Ред.) / Руководство по клинической лабораторной диагностике. М.: Медицина, 1982.-576 с.

Layton D.M. and Mufti G.J. (1987) The Myelodisplastic Syndromes. In: British Medical Journal (1987) Vol. 295, No. 6592, pp. 227-228.

CHAPTER 7

An Overview of the Environmental Problems in the Mining Sector of the Ukraine

Prof. Gennady Averin

Environmental Protection Department
Donetsk National Technical University
58 Artyoma str., Donetsk 8300
Ukraine

Summary

This chapter discusses a set of problems related to environmental safety in the mining sector in the Ukraine, emphasising their regional nature.

Introduction

The problems of environmental safety in the mining sector of the Ukraine are of a regional nature, being heavily dependent on the industrial development of the region. At present, production activities in the Donbass collieries are causing a number of dramatic environmental problems: pollution of surface water bodies by mine effluents, hazardous hydrogeological situations caused by mining operations or mine closures, air pollution caused by burning of spoil piles, stockpiling of industrial wastes, land damage, etc. All this is evidence of the serious environmental hazard presented by collieries. Mining regions need new analytical techniques to assess ecological hazards using system approach methods. The approach could be used as a basis for the development of emergency situation management systems. The Donetsk region is investigated here as an example to outline discussion trends on the principles of long-term policy on environmental safety.

The onrush of the globalization process at the end of the 20th century brought about new safety challenges. New ecological threats of strategic importance have appeared. They are substantially different from clearly felt threats of terrorism, ethnic and regional armed conflicts, corruption, political expansion. Nevertheless, they do present an actual threat to national safety in many countries.

At present the term "environmental hazard" refers mainly to environmental processes such as: climate changes, ozone layer depletion, acid fallout, destruction of rain forests, environmental pollution, biodiversity depletion, etc. Most of the processes are of a global character, though they arise locally or regionally, where control efficiency is very low, which is why it is of crucial importance to provide an adequate analysis, by region or by industry, of how environmental hazards develop.

Economic growth, now beginning to show, and a certain positive stability in the Ukraine are progressing in the face of ever increasing environmental problems and growing man-caused industrial hazards. Moreover, the problems are most urgent for the eastern industrial regions of the Ukraine. Historically, the Donetsk region features a long-established large mining complex, which provides 20% of the national gross domestic product. Industrial production is taking place in most hazardous sectors with the highest technological load per unit area. Man-caused impact on the environment has exceeded European standards. Within the boundaries of the region, about 900

45

W. Leal Filho and I. Butorina (eds.),
Approaches to Handling Environmental Problems in the Mining and Metallurgical Regions, 45–52.
© 2003 *Kluwer Academic Publishers. Printed in the Netherlands.*

large industrial plants include 140 collieries, 40 metallurgical plants, 7 thermal power stations, 177 chemically dangerous operations. Moreover, there are 300 mineral deposits now being mined 1,230 km oil/gas pipelines and ammonia pipelines, about 160 large hydraulic facilities and 113 other operations using radioactive agents. About 3.8 million people, i.e. 80% of the population of the Donetsk region, live in potentially endangered zones (*The Land of Discontent...*).

Mining stands out as the most harmful industry against the industrial background of the region. The situation is increasingly aggravated by the condition of the present mines: of 284 mining plants 100 mines were commissioned forty five years ago and 52 mines have been in operation for about hundred years. Over twenty five per cent of the stationary equipment has exceeded its recommended service life, about half the surface systems have been operating for 40 years without any refurbishing whatsoever and production facilities have deteriorated significantly.

As a result, coal production has fallen as has productivity, while the cost of coal has gone up. The output from mines has more than halved: from 165 mln t in 1990 to 80mln t in 2000. Technologically and economically coal mining appears to be very inefficient: average monthly productivity in coal production is 21.7 t/man, which is substantially lower than the world standards in leading coal-producing countries. One ton of saleable coal with an average ash content of 23% costs 21.5 hryvnias. On the whole the mining sector is detrimental.

The slump in coal production continues despite increased use of coal in the energy sector. Coal-powered electric stations generate approximately half (44%) of the world's electricity. Countries producing electricity from coal include Poland (96%), Denmark (93%), Australia (86%), China (70%), USA (60%), Germany (58%) and others [*Kopylov V. et al.*]. It is estimated that the specific importance of coal in the energy consumption balance will rise in the 21st century. According to the International Energy Agency, world coal trade is expected to double. Experts argue that the demand for steam coal will grow by 2.5% yearly, while the position of coking coal will remain unchanged. Aware of the scarcity of oil/gas deposits in the world, advanced countries are taking measures to develop the coal power sector. The USA, for instance, having far greater supplies of hydrocarbon materials than the Ukraine, have adopted and successfully implemented a 10-15-year special-purpose programme. In this period 275 new coal-powered stations have been built and many of the liquid fuel stations have been switched over to coal.

In the Ukraine with its limited oil/gas resources (supplying only 10-20% of the demand) and vast deposits of various coal grades, coal is the only dependable domestic stock (fuel) both in the short term and in the long term. In 2001 a long-term National Energy Programme (NEP) was adopted. The Program envisages drastic refurbishing of the power industry and outlines a development strategy for the fuel and energy complex up to 2010. It plans an increase in the use of domestic energy resources from 44.4% to 60.8% in 2010.

In the context of the programme, the government of the Ukraine has adopted the *Ukrainian Coal Programme* to develop the coal industry in the period 2001-2010. Aiming at development and strengthening, it involves setting up ten new underground mines and one surface mine with a total output of 23.9 mln t. In addition, ten coal mining operators are planning to put into operation second stage or new production facilities and to raise the output by 17.7 mln t by 2010. During the period 2001-2010, thirty nine mines will be modernized and 38 now working mines will be closed. Total coal production in the Ukraine is planned to reach 170 mln t by 2010. State budgeting of the industry is envisaged at 34 bln hryvnias for the period 2002-2010.

However, *Ukrainian Coal* clearly pays insufficient attention to the ecological problems of the industry. For example, there are modernization and re-equipment plans for about 350 colliery facilities but few environmental protection measures, particularly with regard to cleaning mine waters, are planned. Re-equipment allocations amount to 18.2 bln hryvnias, of which less than 0.04% is assigned for use in environmental protective measures. Only one problem, of many which exist in the industry, is highlighted in the programme for the next ten years – the hydrogeological situation at mines in operation and after closure.

It is obvious from the programme that in the mining industry, now as in the past, money is being saved at the expense of necessary environmental measures. So much the worse for the industry which has the most severe impact on the environment and causes far-reaching changes in nature.

As of today, about 10 bln t of industrial wastes have accumulated in Ukrainian Donbass, with a waste load of 320 thousand t/sq km. The wastes originate from the mining, power and metallurgical industries. In some industrial conurbations and agglomerations the waste load exceeds 1 mln t/km^2 (Donetsk, Makeyevka, Gorlovka), reaching 2.3-3.0 mln t/sq km in others (Enakiyevo, Torez, Dzerzhinsk).

Coal mining with complete caving caused the disturbance of over 600 km^3 rock strata (*Yakovlev Ye. et al*), surface subsidence occurred over an area of 8,000 km^2 (averaging a depth of 1.5-2 m, but in some places, for instance in western Donbass, in the flood lands of the river Samara, down to 7.5 m). Large-scale mining operations with existing mining methods caused marked changes in the geological dynamics of the Donetsk basin. The total mining area covers 11,500 km^2, amounting to 75% of the basin, which itself covers about 15,000 km^2.

Mining operations cause heavy drainage of water-bearing levels. As a result, coal mining is associated with heavy flooding, totalling 25 m^3/s (almost 800 mln. m^3 annually), which is in excess of annual natural river run-off in the Donetsk region. Water emission in mines now reaches 8-10m^3/t affecting the water environment in the region. At an average mineralization of 3kg/m^3 in mine water, about 2.4mln t of salts are carried away. Long-term environmental monitoring shows that underground water mineralization has risen from 0.5-1 to 1.5-3 g/l over the last 30 years due to the increase of man's impact on the hydrosphere. Meanwhile, the amount of underground fresh water (under 1g/l salt content) has declined four times. Waters with increased salt content (1.5-3g/l) can be found in 83% of the region. Mine waters contain phenol, rhodonite, lithium, heavy metals (iron, copper, zinc etc.) (*Yakovlev et al.*)

At the end of 2001, there were 7,088 stationary sources of air pollution discharging about 900 thousand t of harmful agents in the mining sector. Spoil piles pose the greatest threat of contamination. There are now 933 spoil piles, of which 178 are burning. Piles cover an area of 6,220 ha, containing about 2.3 bln m^3 rock. It should be noted that the extinguishing of spoil piles is the most challenging problem which has to be solved by coal mining operators in Ukraine.

Mine closures contribute to environmental deterioration through flooding and changes in the hydrogeological situation in mining areas. In the Donetsk region there are over 100 locations registered as liable to shears, communities in over 50 locations are at risk from flooding due to the rising ground water level. Mine closures highlight the problems of hydrogeological safety and methane emission.

Of no less importance is the social aspect of ever-increasing industrial hazards as these lead to social tension. Notwithstanding the cut-back in production and measures taken to improve safety, the number of accidents at mines does not seem to be decreasing. Coal mines in the Ukraine suffered 27 category 1 accidents between 1990 and 2000. The most serious accidents occurred at Yuzhnodonbasskaya Mine (1991), Sukhodol'skaya–Vostochnaya Mine (1992), Slavyanoserbskaya Mine (1994), Skochinskogo Mine (1998), Barakova Mine (2000), Zasyad'ko Mine (1999, 2000, 2001). Outburst mines and over-category mines are considered the most hazardous, accounting for about 70% of accidents. In recent years, first-category accidents have also been recorded at less dangerous mines (Ukraina Mine, 2002), a fact which can be attributed to relaxed technical requirements.

Man-made environmental impact as a result of mining operations are difficult to eliminate and alleviation a cost-intensive process. Even after closure the mines can still present an ecological threat. Some of the major ecological problems in the industry are:

- lack of practical mechanisms and incentives encouraging environment-friendly production;
- lack of funds to implement environmental measures (cleaning slurry sumps, preventing self-ignition in spoil piles, land rehabilitation etc.);
- environmentally inadequate mining methods, lack of money to introduce low-waste, material- and energy-saving technologies (wastes recycling, low-ash technologies, mine water utilization, coal seam drainage and methane utilization etc), R&D inefficiency of mining operations compared to world standards;
- hazardous hydrogeological situations in working and closed mines;
- increasing contamination of surface water bodies by mineralized mine waters and the lack of economically sound technologies and management to solve the problem;
- falling rate of rehabilitation of impaired lands;
- inefficient extinguishing techniques (20% of piles are burning; despite the reduction of working piles, the number of burning piles has not decreased from 1996-2001);
- no funding of R&D, exploration, servicing, environmental research or analytical laboratories;
- inefficient ecological assessment techniques that require further development;
- outdated monitoring and analytical systems used in mining areas (no updating for the last twenty years);
- weakness of state policy unable to provide environmental and industrial safety for mines, environment management systems provide only inadequate control of water and land resources.

Existing environmental problems at collieries have been accumulating for a long time. They are of a complex nature and feature both technological and socio-economical aspects. Some ecological hazards are obviously industrial and regional. So it is vital that system analysis procedures for ecological assessment of the industry be developed.

The challenge is that up to now there have been no generally accepted criteria for assessing total environmental and man-caused hazards at industrial sites. Although there are assessment criteria [*Directive 96/82/EC*] for operations with hazardous agents, other industries cannot claim generally to have used the criteria needed to assess geological and hydrogeological hazards, or technological and man-caused risks at industrial plants.

Through practice we have acquired some skills and developed certain assessment procedures for various hazards. For instance, in mining the following three major approaches are used: expert evaluation of the facility (parameters); evaluation of the facility (parameters) compared with normal conditions, reference or background values; and evaluation by long-term statistics. At present the work of experts and inspectors with Gosnadzorokhrantruda (Mine Inspectorate) and the Ministry of Environment and Natural Resources of the Ukraine is based on the first approach.

It is known that in large-scale hazardous operations, such as mining, safety expenditure is only loosely linked to the production rate. This implies that, though such operations feature a higher potential hazard, they themselves have the potential to reduce the risks involved (*Marshall V.*). In the context of industrial hazards analysis, collieries hold a special position, as the prediction of the mining environment is very uncertain. This is the reason why expert assessment of hazards in mining, where a specialist is an expert, is crucial. Moreover, the experience of an inspector or engineer is a major factor in valid assessment.

Hazard prediction is usually performed by components, making up certain mine and mining environment. A systematic approach is recommended which will, step by step, predict hazards, assess significance and develop measures to prevent hazardous situations:

- identification of possible hazards in mining processes or facilities;
- investigation of present mine and mining conditions;
- surveying how the standards and current mining regulations are met;
- prediction of hazard levels;
- estimating the significance of possible scenarios;
- development of measures to prevent hazardous situations, accidents and their consequences.

It is known that many negative and dangerous situations in mining can be caused by a number of factors, and primarily by economic difficulties at the plant, obsolete technologies and negative social processes in the country concerned.

In our opinion, it is the sluggishness of technological systems that governs the situation in Donbass: the accident rate is not decreasing while the production rate appreciably is. In advanced countries the number of accidents is more or less stable despite an ever-increasing production rate. Safety experts attribute this fact to a weak link between the production rate and safety mechanisms. Sluggishness of technological development is attributed to the fact that no easily tangible benefits are brought about by any increase in safety expenses [*Marshall V.*]. It has long been practice in the Ukrainian mining sector that material and technical supplies, as well as intellectual resources have been allocated to the development of coal mining and preparation. Less attention has been paid to relevant safety and labour protection measures and even less effort and money has been devoted to finding sound solutions to environmental problems. Thus, environmental problems have been accumulating, industrial impact has been growing and endangering not only the personnel but also adjacent communities and the region as a whole.

Current regulations and standards specify general approaches to the assessment of environmental and industrial impact. For instance, in 1997 Russia adopted the Federal Act on Industrial Safety (*Federal Act*). In the Ukraine a similar act is under consideration. The acts allow for certain provisions from the previously mentioned Directive 96/82/EC, referred to as SEVEZO II. The Directive specifies procedures for assessing the dependability of industrial systems based on evaluating technological and natural risks. This approach is becoming widespread in Europe for evaluating environmental and man-caused hazards in various human activities. SEVEZO II covers many issues, is of an integral character and aimed at accident prevention. It is considered useful as a basis for the management of industrial and environmental risks.

According to the Directive methodology, a serious accident is an accident involving at least one of the following:

- an ignition;
- an explosion or a leakage of harmful agents in excess of permissible limits;
- at least one fatality;
- at least 6 injured employees (outpatient treatment or a one-day stay in hospital);
- damage to community buildings outside the industrial site;
- constant or long-term damage to 0.5 ha of ecologically efficient habitat or more than 10 ha of other habitats including arable lands;
- serious or long-term damage to freshwater and marine habitats;
- contamination of more than 10 km of a river or channel, more than 1ha of a lake or pond, more than 2 ha of a river delta, coastline or open sea;
- severe damage to more than 1 ha water bearing bed or ground waters;
- severe property damage amounting to at least ECU 2 million inside the site of operations or at least ECU 0.5 million outside the site of operations;
- transboundary damage after any accident associated with hazardous agents.

According to domestic classification, a 1st category accident is an accident with the following consequences:

- at least 5 fatalities and more than 10 injured;
- an emission of hazardous materials outside the sanitary protection zone;
- more than ten-fold increase in pollutants concentration in the environment;
- destruction of buildings, structures and facilities at the site, endangering life and health of many employees and non-employees.

Coordination of approaches is absolutely crucial in accident assessment as the data received after an accident and those used to determine hazard indices depend heavily on the ranking criteria adopted in such cases. For example, when the criteria for a classification of an accident specify only half the number of deaths, the accident sample size could be 2-3 times higher. This is why the EC Directive criteria cannot be applied on a one-to-one basis to our situation. Assessment of hazardous situations requires a certain harmonization of approaches.

Typically, hazard analysis is based on statistics and expert evaluation. Table 1 shows expert data for potentially hazardous facilities in the Donetsk region (*Risk Passport…*), as defined by the standards adopted in the Ukraine. If the EC Directive were adopted, the list of Donetsk's potentially hazardous facilities and plants would be twice as long. To identify environmentally and industrially hazardous facilities, standard procedures and identification algorithms based on input data relating to the plant are used in European practice. The data include description of the plant's operation conditions, identification of endangering facilities and processes, description of locations and zones where a dangerous situation or an accident might occur. A system approach to assess a hazardous enterprise involves assessment of individual facilities in terms of process safety, use of hazardous agents, their volumes, physical, chemical and toxicity properties and air/water impact. Besides, various scenarios are prepared, accident probability, scale and severity are predicted. Inspecting bodies, typically overloaded with work and unable to process the plethora of initial data, cannot give an adequate assessment of all plants and operations in the region. In European practice, the data must be recorded in a Safety Report, which is mandatory for all potentially hazardous enterprises. Safety Reports are to be updated every five years. In the Ukraine this issue is being dealt with. However, there are considerable differences in the approaches, with regard to the list of potential hazards and assessment procedures.

Table 1. Characteristics of hazards in the Donetsk region industrial complex

№	Characteristics of facilities and areas	Quantity
1.	Potentially hazardous enterprises and operations	612
2.	Industrially unsafe facilities and areas	48
3.	Environmentally unsafe facilities and areas	18
4.	Operations and facilities associated with utilization and storage of highly hazardous chemical agents	10
5.	Highly hazardous facilities associated with fires, explosions and radiation, including over-category mines	150 50
6.	Cities and districts with over 850 persons/km^2 and potentially hazardous enterprises and operations	50

It should be noted that negative trends of man-caused impact on the environment cannot be eliminated in 10-15 years, and industrial regions are unable to drastically alter the situation.

Conclusions

Despite the rather negative contest, the Donetsk region has certain opportunities to create a well-balanced reasonable regional policy based on systematic and coordinated actions of regional authorities, industrial operators and the public. A way to implement the policy could be to introduce high technologies in some enterprises (*Donetsk Region Programme…*), though the share of the enterprises will be insignificant and appreciable changes will be seen only over a long-term period. In the world economy there are sustainable processes featuring the transfer of energy/material-consuming operations and environmentally unsafe plants to the outskirts of economically advanced systems. This is why the regional economy will for years to come remain biased towards energy-consuming technologies, the power, fuel and metallurgical industries, where environmental hazards are difficult to control. World practice knows only one way to improve safety: enlargement, technical upgrading, re-equipment, rise in productivity and corporate culture.

Along with large enterprises there will be many small operations with primitive technologies and poorly qualified labour. Thus, environmental safety should be oriented to a significant differentiation of industrial enterprises, both in terms of technology and production. Regions and areas with very different types of technological development and man-caused impact will quite possibly appear in industrial regions of the Ukraine. This could aggravate the existing regionalization process. The role of the government in this situation should increase, and it is necessary to activate state control and management of environmental safety. Much attention should be paid to strengthening existing social institutions and developing new ones, to providing access to information and developing environmental monitoring systems since true information is of crucial importance in current conditions.

52

Of no less importance are higher standards of environmental and industrial safety. Hence, it is crucial to adapt national environmental and industrial safety legislation to EC standards, thus contributing to co-operation between the Ukraine and the EC.

References

Directive of EC Council 96/82/EC "On Control of Serious Accidents Associated with Hazardous Agents". – Council of EC. – Geneva, 1996. – 22 p.

The Land of Discontent. Outline of Ecological Situation in Donetsk region. State Department of Ministry of Environmental Safety, Donetsk, 1999. – 102 p.

Kopylov V.A., Khlapenov L.Ye. et al. (2000) Coal Industry Restructuring and It's Social and Environmental Consequences. – Ugol' Ukrainy, № 10, 2000, pp. 3-10.

Marshall V.(1989) Major hazards in chemical industry. – M. Mir. – 671 p.

Risk Passport of Emergency Situations. Donetsk Region. – Donetsk, 2000. – 31 p.

Donetsk Region – 2010 Programme. Oblderzhadministraciya. Donetsk. 1999. – 110 p.

Federal Act of RF № 116 – Ф3 "On Industrial Safety of Hazardous Industrial Facilities" of 21 July, 1997.

Yakovlev Ye.A., Slodnev V.A., Yurkova M.A. (2001) Mine Waters – Ecological and Hydrological Factor in Mining Regions. Ugol' Ukrainy. № 6, 2001, pp. 18-20.

CHAPTER 8
Methods for environmental audit in a mining complex

Prof. Sergey Zenchanka[1], Lakovets Ju.[2],. Zenchenko A.[2], Zhukov N.[1]

[1] Environmental Monitoring Department, International Sakharov Environmental University, 23, Dolgobrodskaya Str., Minsk, 220009
Belarus
[2] JSC "BelEcoMedService", 35, Kujbysheva Str., Minsk, 220029
Belarus

Summary

Conducting the preliminary environmental audit is an important part of enterprise certification in accordance with the International standards ISO 14000 series. An environmental audit of a mining complex is a very complex process involving an environmental audit of both the plant itself and the surrounding region. Mining has a great influence on the environment in terms of pollution of atmospheric air, changing of natural landscapes and existing ecosystems, surface pollution by waste, pollution of underground waters and fresh reservoirs, pollution of top-soil, surface subsidence and so on.

The environmental management system is considered as part of a triad which includes quality management system and social accountability. The application of remote sensing systems and GIS technologies during the preliminary environmental audit is discussed. Considerable attention is paid to the dataware of environmental management systems. Common principles of creation and functioning of such systems have been considered.

Introduction

Taking into account the necessity of maintenance and improvement of quality of an environment, protection of population health, the attention to potential influence of enterprise activity, their production and services on the environment grows. The environmental efficiency of a business becomes of increasing importance for both internal and external interested parties.

The modern requirements of the world economy in the field of the environment, increasingly more stringent legislative requirements, both national and international, infringe on the interests of manufacturers and force them to reflect on their position regarding environmental protection. It is possible to say that this is the basic incentive and motive for the creation at individual plants of an environmental management system on the basis of the standard ISO 14001 (ISO, 1996a).

A series of the standards ISO 14000 was accepted in the Republic of Belarus as the national standard in 1999-2000, though work on the creation and implementation of environmental management systems began in some Belarus plants in 1996-1997. In 2001 the first plants received environmental certificates for the established environmental management systems (Shumilo, 2001; Kuchko, 2001)

W. Leal Filho and I. Butorina (eds.),
Approaches to Handling Environmental Problems in the Mining and Metallurgical Regions, 53–63.
© 2003 *Kluwer Academic Publishers. Printed in the Netherlands.*

Development of the environmental management systems

Systems of environmental management are developed in order to facilitate decisions regarding problems of environmental safety in industry, to give industries a favourable ecological image, to reduce the effect they have on the environment, to satisfy ecological requirements of a community, etc. Standards of environmental management and audit of the series ISO 14000 are now recognised all over the world and successfully implemented in various branches of industry. In European countries the system EMAS (Environmental Management and Audit Scheme) is also employed.

The specified systems can work in plants of various sizes and environmental status, and in various branches. The basic requirement is the realization of a programme of constant improvement and reduction of influence on the environment. The success of the system depends on the obligations taken on at all levels and all divisions of the organisation, especially by more senior managers. Such a system gives organisations an opportunity to establish procedures and to estimate their efficiency, to formulate their environmental policy and environmental objectives, to achieve conformity to this policy and environmental objectives and to show this conformity to others. The general purpose of this standard consists in supporting measures on protection of the environment and pollution prevention whilst preserving the balance with socio-economic requirements. It is necessary to note that many requirements can be considered simultaneously or be reconsidered at any time.

Such a general approach requires environmental auditors and consulting companies to develop specific methodical recommendations for each individual case.

Features of certification in the mining industry

The areas in which mines are located differ in terms of complex combinations of natural and man-caused impacts. Thus, the influence of the latter exceeds the ability of natural forces to reconstruction on power influence, that results in irreversible changes in the geological environment (Lushchik, 1999). As a result there is considerable influence on the environment (industrial and bordering regions):

- Pollution of atmospheric air (discharge of dust and gas from boiler and technological processes, transport, erosion, etc.);
- Changes in natural landscapes and existing ecosystems by man-caused formations (dam, soil-reclamation canal, etc.);
- Surface radioactive pollution;
- Surface pollution by industrial waste;
- Pollution of underground waters and fresh reservoirs, and also other sources of water supply;
- Pollution of topsoil (wind erosion, waste dissolution, etc.);
- Shearing of mountain rocks;
- Occurrence of seismic zones of different scale levels (Malovichko, 1999)
- Subsidence of ground surface (up to 4,5 meters (Tjashkevich, 1999));
- Underflooding and swamping of land;
- Changes in biodiversity (change of species diversity, population, long-term, seasonal and even daily biorhythm of animals (Demidovich, 1999), pollution of woodland)
- Negative influence on environment at test well-boring.

In addition, the industrial complexes present the following environmental aspects:

- Increased power consumption for control of climatic parameters in mines (Krasnoshtein, 1999a);
- Increased power and fresh air consumption for mines aeration (Krasnoshtein, 1999b);
- Dangerous and harmful working conditions of the workers.
- Dangerous epidemic situation.

It is necessary to note that the production of many mines represents the first stage in the life cycle of production of other businesses. This means that more attention should be paid to their influence on the environment. With regard to the environmental certifications of potassium mining complexes, it is possible to specify the environmental danger involved in using some of its products (for potassium fertilizers - increased contents of K^{40} in agricultural soils, formation of physical clay, pollution by accompanying elements, for example, Cl, etc. (Chertko, 1999)).

Thus, specific to the mining industry as a whole is the danger they present both at an environmental level and with regard to conditions in the mines themselves. It is also necessary to consider epidemiological aspects and the status of the environment at the place of work. A specific approach to their certification, including certification of the plant and of the region, is required.

If the mines, or more correctly, their environmental management systems are objects of certification according to the standards of ISO 14000 series (ISO, 1996a), the region is not an object of investigation. Nevertheless, the managing documents of a subsystem of environmental certification in the Republic of Belarus do provide an opportunity to give environmental certification of the regions involved (Managing, 2000).

By preparing the plant and/or the district for environmental certification it is necessary to have a preliminary estimation of their environmental state. In Russia such a stage is based on the Order of the Ministry of Natural Resources from July 8, 1998 № 168 'On the realization of an inventory of licenses on the right of use of the bowels of the earth' and Order of the Minister of Natural resources from April 2, 1998 № 95 'On the creation of system of audit of the use of the bowels of the Earth', which determine the license conditions on the use of the bowels of the earth and basic measures on the establishment, development and introduction of an audit of the bowels of the earth (Tjurjukanov, 1999). In the Republic of Belarus this stage has been given the name 'preliminary environmental analysis' (Environmental, 2000). We will use term "preliminary environmental audit" in this chapter.

Though the preliminary environmental audit has not yet been fully developed, it is expedient to define common criteria and rules on which it will be carried out. For introduction into the world economic system it is expedient to be guided by the standards ISO 14010 and ISO 14011 (ISO, 1996b, 1996c). As the task of maintenance of environmental safety is indissolubly connected with maintaining the quality of production, when performing the preliminary environmental analysis it is necessary to use as a basis results received when quality management systems according to the standards of ISO 9000 (Okrepilov, 1998) were introduced in the plants concerned.

Analysis of social aspects

Usually, evaluation of the influence an enterprise has on an environment is expressed in quantitative terms, such as emissions, sewage disposal, waste, etc. Frequently these data are considered together with results of epidemiological research that allow us to estimate the danger of the plant to health and lifetime of the population. Nevertheless, when performing the preliminary environmental

56

analysis it is expedient to consider a complexity of social aspects: living conditions, social structure, urbanization of territory, etc., and not just data from medical and epidemiological research. The conditions at workplaces should correspond to the standard SA 8000 (International, 1997), at present not accepted in the former Soviet Union countries. This data cannot be considered separately from the economic problems of the enterprises and regions.

Basic stages of realization of the preliminary environmental audit

As the basic tasks of the preliminary environmental audit it is necessary to consider:

- Estimation of influence of the plant on the environment;
- Subsequent details of environmental policy;
- Establishment of environmental objectives and targets;
- Creation of information base for development and introduction of the environmental management system at the respective plant;
- Estimation of a professional level and awareness of the necessity for environmental protection among all levels of staff in all departments.
- Opinions of the interested parties on the ecological activity of the organisation.

For the decisions concerning the problems posed the preliminary environmental audit of the enterprise should include (Environmental, 2000):

- Estimation of compliance by the plant to the requirements of the legislative and normative acts including licensing of separate kinds of activity;
- Rating and registration of all practically significant influences on the environment and definition of the most significant of them;
- Examination of existing management practice in the field of protection of the environment and use of natural resources;
- Analysis of the previous facts pertaining to ecological incidents and discrepancies and measures taken to eliminate them.

These stages practically coincide with the stages of an environmental audit spent at certification of environmental management systems and serve to reveal environmental aspects and the development of the enterprise's environmental policy.

In the analysis it is necessary to consider all regular and non-regular modes of operation and possible emergencies irrespective of their cause.

As mentioned above, when rating the status of resources on the site of and bordering on a mining complex, it is necessary to estimate the intensity of land use, land erosion, underflooding, subsidence of rock, withdrawal of grounds under career, waste banks, landslides and waste dumps, chemical soil pollution, soil pickling and acidation, dehumusing of grounds, pollution of superficial and earth waters, etc. Thus, the rating of an area status also includes a rating of the water systems. It is also necessary to describe existing and potential influences on fauna and flora. Field observation is employed only to an insufficient extent in carrying out these tasks.

Use of remote methods and GIS technologies in carrying out the preliminary environmental audit

The scale of influence the mining operation has on an environment requires a preliminary environmental audit followed by the application of modern methods to estimate the status of the environment based on results of remote sensing and GIS-technologies. These methods should not be applied alone but only in combination with procedures of field observation, such as reconnaissance survey and instrumental inspection.

Now there are plenty of satellite systems with various spectral (up to 50 spectral channels) and resolution (from 2-3 up to 1000 and more meters) characteristics (NOAA, Resource, Terra, SPOT, LANDSAT etc.). The air systems for remote sensing, as a rule, have both higher resolution and higher optical distortions and complexities at processing.

These systems are applied to forecasting and evaluation of resources, economic evaluation of various natural resources, environmental assessment of land and objects, assessment of environmental and economic loss from nature management, account of dynamics of every possible natural and human impact etc. There are both direct and oblique methods of definition of necessary parameters of sensing territory.

The processing of remote sensing data is a special area of images and spectrum processing. There is a separate class of specialised software for the processing of remote sensing data is clearly distinguished from systems used for processing images of general purpose. The initial information is the image, spectral image and polarizing characteristics of the object submitted in a digital format. Digitalization is carried out at the board of the flying object and further data are transferred to the computer of the stationary processing centre for preliminary and further thematic processing.

In modern systems sufficiently precise binding of the raster images of remote sensing is possible through the system of global positioning GPS (for example, at air sounding), and also through relaying the orbit of the flight of the satellite to geo information systems (GIS), which makes it possible to carry out various mathematical and statistical accounts with a high degree of accuracy.

The combination of technologies of remote sensing and GIS allows us to solve a huge quantity of tasks from the retrospective analysis of various natural objects to forecasting and obtaining exact forecasts of process dynamics. It makes possible the application of these technologies in such areas as an assessment of influence on an environment during the object designing phase, in environmental audit and environmental certification of land and the environmental management system. These methods are especially important for plants working with natural resources.

The combination of remote methods and methods of field observation makes it possible to create a set of GIS-maps covering various themes, for example, soil maps, map of waste bank, etc., for the area in which the plant is located.

In the Republic of Belarus and on all former Soviet Union territory scientific research is being conducted in the field of using the space information and GIS data for monitoring the environment of mining complexes.

For example, in the Ukrainian State Institute of Mineral Resources long-term research (with forecasting elements) is being carried out aimed at an assessment of the geological environment of the regions subject to the most severe and prolonged influence of industrial exploitation. A technique is being developed enabling complex research of geological environments which will make it possible to obtain data on their status at the lowest possible cost and more operatively (Lushchik, 1999).

The first evaluation stage is the interpretation on different scales and at different times of snapshots which reveal sites of development and activization of exogenous geological processes (landslips, landslides, slamps, underfloodings, erosion, drainage, etc.). The following stage is reconnaissance well-boring, testing hydro-geological works, etc. In a final stage cartographical and mathematical methods of research are applied.

As a result of such assessment the necessary information for forecasting the development of the pollution of the top layers of the lithosphere, underground and surface waters, modern dangerous geological processes, including earthquakes, etc., is obtained which makes it possible to use similar methods during the preliminary environmental audit of the plant and land area of mining industries.

A package of complex interpretation of geological-geochemical information « Gold Digger » developed in the Department of Geochemistry in the Faculty of Geology at Moscow State University (Vorobjev, 2002) serves as an additional example.

The package is intended for the processing of the data received by geochemical searches for ore deposits, geological-geochemical mapping of rock and environmental monitoring of industrial regions. It includes the programmes of allocation, classification and interpretation of anomalies at all kinds of ecological-geochemical works. The data processing can be carried out taking into account all the qualitative attributes which the researcher considers it expedient to retain from field observation. The results can be mapped on a raster and vector basis: geological and landscape maps, aerial photographs and so on.

In the Republic of Belarus similar scientific research is being conducted. The systems of remote sensing are used by a state hydro meteorological service, the Ministry of Forestry and the Committee on extreme situations.

Centre "Ecomir" uses GIS for the Soligorsk district – a set of base digital maps and the attributive data on the status of the environment adhering to them (Kovalev,1999). Other maps used are: an aerolandscape; geobotanical; geomorphological; overburden; modern geodynamics; soil; stability of a relief to technical influence; rational use of land, protected natural territories and recreational zones, genetic types top water horizon, hydrochemical; bogs and waterlogged grounds; superficial pollution by Cs^{137}; exposition doze; superficial pollution by K^{40}; natural contents of chemical elements; pollution by emissions the industrial works.

Combining the given information base with the data from space monitoring and verifying it with the data from field observation it is possible to receive the relevant information on changes in an environment and to carry out monitoring of the environment. Using the data from this monitoring it is possible to carry out the preliminary environmental audit and to give the recommendation and to promote acceptance of the administrative decisions.

There is also scientific research on the creation and realization of the connected space geoecological monitoring of the Soligorsk district being conducted at "«Kosmoaerogeology" and Belarus State University. The system consists of the following: space, topogeodesical, meteorological, structural geomorphological, morphometrical, hydrological, hydro-geological, geodynamic, geomechanical, seismological, geophysical, geochemical, mount-ecological (Tjashkevich, 1999).

The space block includes processing and analysis of the space information, retrospective analysis, thematic map-making on the basis of snapshots processing.

The topogeodesical block provides application GPS system for the study of modern vertical and horizontal movements of the ground surface of natural cause or caused by humans. The meteorological block includes the collection, analysis and coordination of the meteorological data in a complex with the data of the retrospective analysis of cloud on the basis of space snapshots. This enables a prediction of the area distribution of atmospheric emissions on the basis of mathematical models.

The structural geomorphological block is based on the fact that, using exogenous geomorphological processes, it is possible to obtain information on endogenous processes and vice versa.

The morphometrical block includes the detailed analysis and dynamics of a relief, its modern natural deformations based on the construction of a number of thematic maps. The hydrological block will capture works on retrospective analysis, estimating status and dynamics of open reservoirs, river and soil reclamation canal. The hydro-geological block provides an establishment of dynamic laws, forecast and chemism of underground waters within the limits of a working regime network.The geodynamic block. The maximal information on tectonics, neotectonics, structural surfaces of the basic marking horizons and capacities of adjournment, data of structural geomorphology and modern natural deformations of a relief and others for maintenance of conditions of safety and minimisation of ecological press on an environment will be used.

The geomechanical block means creation of databases on all mine fields for the construction of a map of the actual isoline of surface subsidence. The seismological block provides the construction of a map of seismic zoning of the Soligorsk district and adjoining regions and an area forecast of seismicity. The geophysical block enables the analysis of all available geophysical material and, whenever possible, reinterpretation of results. In the geochemical block, it is necessary to study archival material, to establish geochemical anomalies etc. The mount-ecological block is on progress. The estimation which has been carried out in this block actually is the preliminary environmental audit for the grounds of the mining plant.

Information system for environmental management of the mining plant

Systems for monitoring and measuring environmental conditions which are suitable for carrying out the preliminary ecological analysis involve work with huge amounts of information. The environmental management system (EMS) is also informative enough. All these factors make it necessary to create a computer information system for the environmental management of mining works. The development of an information system for the EMS of the industry is being conducted in ISEU, Minsk (Zhukov, 2002). The common principles of an information system for EMS:

1. The environmental management system is sufficiently large, and it should be employed at various points in the mining complex. The department of ecology or the company's ecologist is not able to supervise the correct functioning of the system alone. Therefore, the software for the environment management system should also consist of several programmes installed in various

departments of the company. These programmes should be inter-connected and should receive the necessary information from one source (client-server database).

2. The importance of an environmental problem should be known to everyone at the plant, therefore the EMS should function from the highest administrative level (the head, the director etc.) down to the lowest (the master, the chief of department etc.). Besides this, ordinary workers should be acquainted with the politics of the company, they should accept these policies and be responsible for their implementation. The planning of activity in the field of ecology, stating the environmental objectives and targets is carried out according to the ecological policies. Hence, the computer system should have an advanced system of notification and information output regarding changes in ecological policies, (objectives and targets), audit findings, quality and efficiency of EMS functioning. This information should be accessible to all workers in the company.

3. Often the real data on the functioning of the system is not reflected in documentation. The reasons for this are either that the company's documentation system complicates documentation management and allows the documentation to be ignored or that it gives too wide access to the documentation. Proceeding from the aforesaid, the computer system should provide fast and relevant access to the information, but should exclude casual access, processing of the information by inept workers or by workers, whose duties and rights do not include such processing (i.e. there should be a system of passwords and access protection).

In Fig. 1 the information system for environment management is presented schematically.

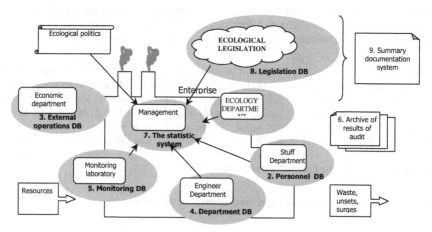

Fig.1. The structure of software for EMS

The common principles of information system construction are applicable to mining operations. However, the certification of the premises of the enterprise is also necessary. Consequently it is necessary to develop an information system for managing the area covered by a mining complex.

Both these systems should be closely connected, and the latter should provide the opportunity to work with graphic information (cuts, structures, aero snapshots, space snapshots, vector maps etc.), statistical and other information. These features present the user with an opportunity for a qualitative visual understanding of the contained data. It is also necessary to provide the interface of the given information system with other software packages used at work, with vector maps and attribute databases, aero snapshots, space snapshots, statistical reports etc.

Schematically such a system is represented in figure 2.

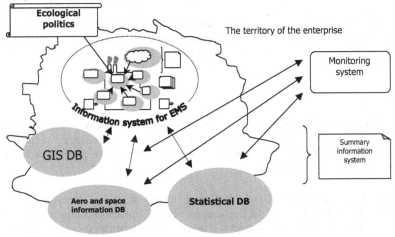

Fig.2. The structure of information system for EMS and for territory management

Such an information system makes it possible for mining industry plants:

- to manage the EMS documentation at a high standard;
- to keep and to use effectively the data of monitoring etc. for the current analysis and subsequent audits;
- to standardize the preliminary ecological analysis;
- to visualize and to optimise environmental management.

Conclusions

The realization of the preliminary environmental audit of the operations of a mining complex is a complex task, in which the economic, ecological and social problems form a triad requiring consecutive and joint decisions which can be made by application of the international standards of series ISO 9000 and ISO 14000, and also standard SA 8000. Taking into account the fact that the standards of the series ISO 14000 are most universal and can be applied not only to the enterprises, but also to the grounds, a preliminary environmental audit is the tool capable of defining the most critical points. The sequence of actions during such an audit should be:

- field observation of the operation including reconnaissance survey and instrumental inspection;
- revealing the environmental aspects of the plant's activities;
- inspection of the area covered by the plant and of the adjacent area using remote sensing and GIS technologies;
- revealing the most significant environmental aspects of the plant's activity on the adjacent area;
- comparison of results of field and remote observations;

62

- revealing the most significant aspects of the influence of the plant on its employees and on the population as a whole;
- analysis of social structure for areas where building is planned;
- analysis of conformity of working conditions to the requirements of the standard SA 8000.

Carrying out the preliminary environmental audit in such detail and applying modern means in so doing is a complex task but such detail will make it possible to estimate the environmental status of the enterprise and the adjacent region and assess its influence on the workers and on the population as a whole.

References

Chertko N.K. (1999) Ecological and geochemical problems of using potash fertilizers on agriculture landscapes of Belarus. The report theses of the International scientific practical conference "Potash salt in Belarus: a status of deposits development, prospect of development, problems" pp. 164 – 166. Minsk.

Demidovich L.A., Antipin E.B., Gledko U.A. et al (1999) Influence of a Solygorsky complex on geosystems. The report theses of the International scientific practical conference "Potash salt in Belarus: a status of deposits development, prospect of development, problems" pp. 150 – 152. Minsk.

Environmental management system. Realization of the preliminary ecological analysis. The methodical instructions. Minsk: Ministry of nature resources. - 2000.

International Standard ISO 14001 (1996a) Environmental Management Systems: Specifications with Guidance for Use (Geneva: ISO).

International Standard ISO 14010 (1996b) Guidelines for environmental auditing. General Principles (Geneva: ISO).

International Standard ISO 14011 (1996c) Guidelines for environmental auditing. Audit procedures. Auditing of environmental management systems (Geneva: ISO).

International Standard SA 8000:1997. Social accountability.

Kovalev A.A. and Kuzmin V.N. (1999) Monitoring of an environment condition of Solygorsk area on a basis of GIS technologies. The report theses of the International scientific practical conference "Potash salt in Belarus: a status of deposits development, prospect of development, problems" pp. 137 – 139. Minsk.

Krasnoshtein A.E., Kazakov B.P. (1999a) Energy saving technology of preparation of air for mines. Report of the International scientific conference "Potash salt in Belarus: a status of deposits development, prospect of development, problems" pp. 133–135. Minsk.

Krasnoshtein A.E., Minin V.V. and Alymenko N.I. (1999b) Energy saving, supporting conditions of safe conducting mountain works. The report theses of the International scientific practical conference "Potash salt in Belarus: a status of deposits development, prospect of development, problems" pp. 123 – 125. Minsk.

Kuchko G.V. (2001) International recognition of the Environment Management System at Minsk Roll Plant. News. Standardization and certification. 2001. № 4, p. 58.

Lushchic A.V., Davidenko I.P.and Shvyrlo N.I. (1999) The basic rules of a rating of a status of geological environment in mining enterprise areas. The report theses of the International scientific practical conference "Potash salt in Belarus: a status of deposits development, prospect of development, problems" p. 136 . Minsk.

Malovichko A.A. and Butyrin P.G. (1999) Seismological monitoring on mines of deposit of potash salts Verhnekamsky. The report theses of the International scientific practical conference "Potash salt in Belarus: a status of deposits development, prospect of development, problems" p.105. Minsk.

Managing documents of the Republic of Belarus 03810.5. National system of certification of the Republic of Belarus. Subsystem of environmental certification. Minsk, 2002, 282 p.

Okrepilov V.V. Quality Management, Moscow.: «Publicher «Economica». 1998 – 639 p.

Shumilo V.S., Gerkis O.P. (2001) Ecological certificate N1. News. Standardization and certification 2001. № 4, pp. 55 – 57.

Tjashkevich I.A., Rahmanov S.K.and Ponaryadov V.V. (1999) Bases of creation and realization of the connected space geoecological monitoring of Soligorsky enterprise area. The report theses of the International scientific practical conference "Potash salt in Belarus: a status of deposits development, prospect of development, problems" pp. 145-150. Minsk.

Tjurjukanov A.A., Seleznev P.V. and Shevchuk A.V. About the development of natural resources users audit. Bulletin " Using and protection of Russia's natural resources ". 1999. - № 7 - 8. pp. 57-63.

Vorobjev C.A. (2002) Geoinformation system for ecological monitoring data processing. Materials of 3rd Interuniversity youth scientist conference «School of ecological geology and rational resource using» pp. 113 – 119. St Petersburg.

Zhukov N.O., Zenchenko A.S., and Zenchanka S.A. (2002) The software for EMS. News. Standardization and certification 2002. № 2, pp. 61 – 64.

CHAPTER 9

Life Cycle Assessment
- implementation and its financial aspects in the Polish mining industry

Małgorzata Góralczyk

Polish Academy of Sciences
Mineral and Energy Economy Research Institute
ul. Wybickiego 7, 31-261 Kraków
Poland

Summary

LCA (Life Cycle Assessment) is a new tool for environmental management. The most important property of the LCA is the „cradle-to-grave" perspective as the analysis focuses on a single product - from raw materials extraction to the final disposal of wastes. LCA associates the energy and materials that go into the studied system with the wastes and pollution it generates. *System* represents a set of single operations related to a product. LCA is a tool that many producers in the world could use in decision-making processes and which can lead to a reduction in the amount of waste produced or to technology changes. Besides obvious environmental advantages, the reduction of negative environmental impact improves a producer's image and competitiveness on the world market. In the mining industry environmental aspects are very important, as all mining and metallurgical activities influence the environment to a substantial extent.

Introduction

LCA (Life Cycle Assessment) is an environmental management tool of increasing importance. Fundamental concepts of LCA, its methodology and application are described now in the ISO 14040-1404X series of standards. These standards have already been introduced in some countries, but in Poland they are at the stage of verification into Polish. However, there are some companies in Poland that are already trying to foresee how, according to the LCA technique, their products influence the environment and what technological improvements are possible in the existing production process in order to minimise the environmental impact.

The most important feature of LCA is that analysis encompasses the whole life cycle of a product or service. This is done in several stages. The first one is the definition of goals and the scope of analysis. Defining the scope involves determining the boundaries of the system. In the widest perspective the boundaries can cover the whole life of a product. In the first phase the functional unit is also defined. The functional unit is the smallest analysed part of the entire system - it can be a product, an equipment, service or process. In the second phase the input and output inventory is created. Inventory consists of all materials and energy that enter the system as well as the outputs of the system such as semi-products, wastes, emissions and pollution. The amount of data collected at this phase is immense; so all outputs are divided into impact categories which represent the recognised environmental impacts – ozone layer depletion, green house effect, etc. The next step in LCA analysis is data interpretation – here the aspects of the production process which place the greatest burden on the environment are identified.

W. Leal Filho and I. Butorina (eds.),
Approaches to Handling Environmental Problems in the Mining and Metallurgical Regions, 65–71.
© 2003 *Kluwer Academic Publishers. Printed in the Netherlands.*

In the mining industry the environmental aspects are very important, as all mining and metallurgical activities influence the environment to a substantial extent, especially by generating a massive amount of solid waste. The primary non–ferrous industries are a particular example of this, as non–ferrous ores contains only a few percent of metals. The rest of the extracted mineral has to be eliminated gradually in the successive stages of production.

LCA in Polish mining industry

The basic metals sector is a significant component of Polish industry. It contributed only about 5% of the sold production of industry, but the export of metals and metal articles have contributed about 15% of export revenue in the last ten years. Metalliferous mining and processing are important, with copper and its various by-products (silver, gold, platinum and palladium, lead, cobalt), lead, and zinc being the main commodities produced. Zinc and lead deposits are nearly exhausted but copper deposits will allow copper mining for at least another 25 years.

In Poland one of the main assumptions for minerals industry policy - introduced in 1989 - is the pro-ecological development of the industry. As a result there have been a lot of legal changes, of which one of the most important was the introduction of high fees and penalties associated with polluting the natural environment, e.g. fees for emission of dusts and gases, for storing wastes. In order to reduce environmental penalties many companies started to invest mainly in pro-ecological operations, and now emissions to air and water mainly from non-ferrous mining industries have declined dramatically (Table 1).

Table 1. Environmental fees paid by KGHM Polska Miedź SA (Europe's largest primary copper producers)

YEARS	1996	1997	1998	1999
Investments in environmental protection ['000 $]	16 286	41 280	6 380	7 520
Environmental fee [$/1t Cu]	96.24	53.94	38.78	32.79
Environmental fees ['000 $]	17 765	19 495	15 123	15 274

Source: KGHM Polska Miedź S.A.

The other result of the Polish economic reform was the increase in the power price, which also caused a reduction of energy consumption for metals production as well as for other materials (Table 2).

These technological changes, which were necessary from an economic point of view, influenced environmental effects. As a result some mining companies started to apply for ISO 14000 certificate (e.g. Cedynia Copper Wire Mill), and to evaluate their influence on the environment according to LCA criteria.

For new mining projects, concentrating to some extent on the protection of the environment, implementation of the LCA is possible only by using already existing databases. The data from previous mining operations can be applied and the minimisation of environmental impact can be achieved from the very beginning. Data from ongoing operations provides an opportunity to conduct a real assessment of environmental impact. (Such data are analysed in the EU project: Life Cycle Assessment of mining projects for waste minimisation and long term control of rehabilitated sites, with Imperial College as coordinator).

Table 2. The volume of inputs (materials, energy) and output (pollutants) for Polish copper and zinc producers

YEARS	1996	1997	1998	1999
Copper				
Production of copper [t]	424 708	440 640	447 000	470 494
Material consumption [ore/t Cu]	61.15	56.11	58.4	57.4
Fuel and energy consumption [MJ/t Cu]	10 582	10 391	9 343	9 295
- of which electricity [kWh/t Cu]	1 022	1 163	1 219	988
Water consumption during copper production [dcm3/t Cu]:				
Salted water released [m³/day]	73 235	72 378	59 342	42 989
Dust and gas emission [t/year]	2199	1691	958	977,4
Solid wastes ['000 t]	25 657	24 447	26 202	26 973
- of which utilized ['000 t]	7 024	12 816	18 099	19 887
Zinc				
Zinc production [t]	163 100	170 600	174 800	177 000
Material consumption [concentrate/t Zn]	1.13	1,05	1,16	1,09
Fuel and energy consumption [MJ/t Zn]	16 791	16 421	16 389	15 761
- of which electricity [kWh/t Zn]	3 655	3 697	3 775	3 739

Source: KGHM Polska Miedź SA, ZGH Bolesław, Trzebionka SA., GUS

Mining production as a whole involves a number of different technological stages. Hence, environmental problems should be identified for each stage. The most important environmental problems of underground non-ferrous metals production are presented in Table 3.

Table 3. Environmental aspects at each stage of primary metal production.

Stage	Environmental aspects
Prospecting for raw material	Small local pollution
Mining	Mining water discharge – also salted water
Processing	Solid wastes disposal
Manufacturing	Dust and gas emission and solid wastes disposal
Transportation	Dust and gas emission
Distribution	Dust and gas emission
Use	—
Waste utilization	Solid wastes disposal
Recycling or final disposal	Improvement of the environment

The basic analysis of Life Cycle Inventory for Polish non-ferrous mining (by comparing the inputs, i.e. energy, materials, and outputs, i.e. wastes, pollution) shows that the greatest amount of waste and pollution (also the highest fee) is generated at the mining and processing stage. Therefore, these two stages are further divided into operations, processes and activities to create the database of wastes and pollutants, and to evaluate the significance of potential environmental impacts during any operation or activity (using the results of the life cycle impact analysis, i.e. for every functional unit). In general, this process shows where, in the whole life of the production process, the highest amount of wastes and pollution is generated and assesses the effects on the environment (Life Cycle Impact Analysis). Solutions concerning minimisation of generated wastes and pollution can then be suggested in order to help decision–makers in their work (Life Cycle Improvement Analysis).

Solid waste disposal is now the largest environmental problem for **KGHM** Polska Miedź SA, which has been producing copper for 30 years. The capacity of existing tailing ponds can be sufficient for future production (the mine life based on the reserves available for extraction has been calculated for the next 25 years), but the environmental hazard and the cost of reclamation and monitoring (in accordance with Polish law) will have to be conducted for the 30 years after closure. To solve the problem of post–flotation waste management, research has been conducted into the possibility of utilising post–flotation wastes in underground mining technology to backfill exploited areas and to fill abandoned works. Probably, such a solution – at present too expensive – will be the appropriate one after implementation of LCA, and will make it possible to minimise disposed solid and, at the same time, the most hazardous wastes. It will then reduce the amount of fees and fines that KGHM has to pay for the generation of wastes, and enable them to avoid the cost of tailing damp reclamation and monitoring after closure of the mines.

Financial aspects of LCA

Life Cycle Assessment (LCA) is a tool that can help producers make better decisions pertaining to environmental protection. LCA deals with the environmental impact of a process or product during its entire life, but it does not take account of the financial aspects. However, it can be a valuable additional factor in investment decisions when basic investment evaluation methods fail to give an unequivocal answer. This can occur when competing projects are found to be financially equivalent.

Even so, for every project under consideration it is essential to perform a normal investment evaluation, since – as with all decisions that involve money – the financial aspects are crucial. If the project is unsatisfactory in financial terms it cannot be implemented, even if this means rejecting solutions that are environmentally safer. One of the reasons for this is that the company's shareholders expect to earn a profit on the money they have invested. They will not accept a loss or break-even situation in the long term, and if such a situation should take place they will withdraw their funds from the company and invest them elsewhere. Therefore, every environmental investment should be screened with the usual investment evaluation methods. There are several methods used to evaluate investments. The most popular are NPV (Net Present Value) and IRR (Internal Rate of Return). NPV is the value of the project expressed in the first year of project life. It is calculated by discounting all costs and incomes during project life using the following formula:

$$NPV = \sum_{i=0}^{n} \frac{CF_n}{(1+i)^n} \quad \text{where CF – cash flow, i – interest rate, n – number of years.}$$

A positive NPV indicates that, by this criterion, the project is profitable. Another criterion, the IRR, is a rate of return that would make the NPV equal to zero, which indicates the minimum rate of return that can be accepted in order to achieve a profit. One typical problem encountered in such estimates is how to determine the annual cost of the equipment purchased. Example 1 illustrates how to make such calculations.

Example 1. Annual cost calculation[1]

An underground mine intends to purchase a drill car for $40,000, with a lifetime of 10 years and a salvage value of $10,000. The minimum rate of return on this investment is set at 15% per year. The following equation can be used to determine the annual costs for this drill car:

$$\text{Annual cost} = C\left[\frac{i(1+i)^n}{(1+i)^n - 1}\right] - L\left[\frac{i}{(1+i)^n - 1}\right]$$

where C – initial cost, L – salvage value, i – interest rate, n – number of years.

The first component of this equation is equal to a uniform series of end-of-period payments equivalent to a present lump sum of money. The second component indicates a uniform series of equal end-of-period payments equivalent to a future sum of money. In this example the annual cost for the drill car will be $7,478 ($40,000 x 0.1993 – $10,000 x 0.0493).

The calculation of annual costs is very important because project costs and revenues must be known in order to calculate the NPV and IRR. It is also important to include in the project plan the costs and savings that pertain to environmental protection. Apart from the initial investment, these costs should cover technology (usually more expensive, but better for the environment), pollution monitoring, rehabilitation costs, final disposal costs, and energy and maintenance costs, which are often neglected. The cost of performing LCA should also be taken into account. As for savings: these include not only financially measurable items, such as reduced (or even eliminated) fines for pollution, and lower operation and maintenance costs, but also non-measurable effects, such as improved company image and increased competitiveness. In LCA, particular functional units are analysed in terms of their environmental input and output, but the financial implications are omitted. However, input and output can be measured for each functional unit in monetary terms.

Even such output as contribution to the greenhouse effect can be monetised in terms of fines that will have to be paid for emissions. Cost analysis should be performed for each functional unit. Depending on the product or process under analysis, it is crucial to define the functional unit accordingly. The functional unit can be defined as one machine, for example an underground mine drill car with a 10-year life span and a purchase cost of $40,000 (including freight and installation). The primary input and output of such a functional unit are presented in Table 4, and selected costs associated with it are presented in Table 5.

Table 4. Functional unit inventory

Input		Output	
Fuel [kg/h]	24.56	CO [g/h]	2.80
Electric power [kWh]	50	NO_x [g/h]	8.20
Lubricants [g/h]	0.30	Lubricants used [g/h]	0.30

[1] Stermole, F.J. and Stermole, J.M. (2000) Economic evaluation and investment decision methods. Investment Evaluation Corporation. Colorado

Table 5. Functional unit cost estimation

Cost type	Amount
Purchase (incl. freight and installation) [$]	40,000.00
Power (fuel and electric) [$/h]	15.00
Lubricants [$/h]	0.04
Operator wages [$/h]	5.00
Maintenance labour costs [$/h]	5.00
Fines for CO emission [$/10,000h]	0.63
Fines for NO_x emission [$/10,000h]	6.76
Fees for used lubricant disposal [$/10,000h]	0.0117

For companies already in business, the most common difficult decision involves the choice between existing and new technology. Example 2 shows how to deal with such a dilemma. It is quite difficult to assess the exact amount of income produced by a capital improvement that provides a service (such as a drill car in this example). Incremental analysis is thus required in order to evaluate service-producing investment alternatives that involve initial costs. This analysis is based on subtracting the investment that would require less capital from the one that would place a larger demand on funds[2]. As a result, a cash flow diagram is created with an initial investment that will produce savings and/or revenues in the future.

Example 2. The choice between existing equipment and its replacement

An underground mine is considering whether to purchase a new drill car or to keep the existing one. The installation of a new drill car would reduce operating costs (OC) from $27,000 to $20,000 in year one, from $30,000 to $22,000 in year two, from $33,000 to $24,000 in year three, and from $37,000 to $27,000 in year four. Because a new drill car is more environment-friendly, environmental fees (E) will be reduced from $3,000 to $2,000 each year. The purchase cost (C) of a new drill car is $20,000 and the expected salvage value (L) after four years is $5,000. The minimum acceptable rate of return is 20%.

New drill car

Costs [$ '000]	C = 20	OC = 20	OC = 22	OC = 24	OC = 27	
		E = 2	E = 2	E = 2	E = 2	L = 5
Year	0	1	2	3	4	

Existing drill car

Costs [$ '000]	C = 0	OC = 27	OC = 30	OC = 33	OC = 37	
		E = 3	E = 3	E = 3	E = 3	L = 0
Year	0	1	2	3	4	

---

[2] Stermole,F.J. and Stermole, J.M. (2000) Economic evaluation and investment decision methods. Investment Evaluation Corporation. Colorado

New versus existing drill car

Costs [$ '000]	C = 20	OC = -8	OC = -9	OC = -10	OC = -11	L = 5
Year	0	1	2	3	4	

The incremental analysis with calculation of the incremental NPV and IRR is based upon the comparison between the new and existing drill car.

$$NPV = \frac{-20}{(1+0.2)^0} + \frac{8}{(1+0.2)^1} + \frac{9}{(1+0.2)^2} + \frac{10}{(1+0.2)^3} + \frac{16}{(1+0.2)^4} = 6.4 > 0,$$

The incremental NPV of the project is greater than 0, so revenues can be expected from the investment. In this example, the revenues consist in the savings that will be achieved as a result of purchasing a new drill car. By interpolation the IRR is equal to 34.6%, which is greater than the 20% minimum rate of return, and is therefore acceptable. Both the NPV and the IRR of this project indicate that investing in a new drill car is a satisfactory alternative.

Conclusions

The same method of investment evaluation can be applied to clarify the decision whether to improve existing technology or to introduce a new system. The choice between these two alternatives is constrained by limited funds, resources, and other factors, such as environmental issues. If the two investment opportunities were economically comparable, an LCA would be useful. It can also happen that during the whole life of a product the greatest environmental impact occurs beyond the producer's interest (this mainly pertains to the energy used by product). In such circumstances it is even harder to introduce an environmentally safer product because it does not produce direct profits, and thus companies have no incentive to launch such a product. However, there is a trend among producers to minimise the environmental impact of energy production, so there may be some pressure on them to focus on less energy-consuming products. This undertaking can be supported by LCA, but LCA itself has to be supplemented by proper financial analysis. Common investment decision methods can be used in order to evaluate cost-effective alternatives regardless of the type of functional unit chosen (mining equipment, home appliance, hi-tech equipment, etc.). The preparation of such a broad analysis of an investment ensures that the projects introduced will comply with both economic and ecological standards.

Polish mining producers can expect that implementation of LCA will lead not only to an improvement of the environment, but also to more effective environmental management which, in turn, means cost-saving by reducing wastes emission and reducing fees and fines. It represents the acquisition of an effective tool for decision-making which shows the connection between companies' activities and the devastation of the environment and thus points to means of improving a company's image on the world market.

CHAPTER 10

The tailings dam failures in Maramureş county, Romania and their transboundary impacts on the river systems

Dr. Paul A. Brewer [1], Prof. Mark G. Macklin [1], prof. Dan Balteanu [2], Tom J. Coulthard [1], Dr. Basarab Driga [2], Andy J. Howard [3], Graham Bird[1], Sorin Zaharia [4] and Mihaela Serban[2]

[1] Institute of Geography and Earth Sciences, University of Wales, Aberystwyth, Ceredigion, SY23 3DB, UK

[2] Institute of Geography, The Romanian Academy, 12 Dimitrie Racovita, Sector 2, RO-70307, Bucharest, Romania.

[3] School of Geography, University of Leeds, Leeds LS2 9JT, UK

[4] ARIS Design Institute, Str. Ghe., Marinescu 2/15, Baia Mare, Maramureş County, Romania

Summary

The tailings dam failures that occurred in January and March 2000 in Maramureş County northwest Romania, discharged contaminated water and sediment directly into river systems that drain into the Tisa River, a major tributary of the River Danube. To ascertain the long-term impacts of these spills, a survey of metal contamination in surface water, river channel and floodplain sediment in the rivers of Maramureş County was carried out in July 2000. The results show that over 20% of surface water samples exceed imperative values for Zn, Cu and Cd, and over 40% of river channel and floodplain sediment samples exceed Pb and Zn intervention values. Zn concentrations rise immediately downstream of mine sites and tailings ponds but then fall below EC imperative / intervention values typically within 10-60 km. The long-term environmental impact of the Bozanta-Aural and Novat-Rosu mine tailings pond failures is still difficult to quantify, but their impacts should be seen in the context of adding to contamination arising from decades of poorly regulated, and largely untreated, industrial, mining and urban discharges into local rivers.

Introduction

Since 1970 there have been 59 reported major mine tailings dam failures around the world; in 2000 alone there were a total of 5 reported accidents, two of which were in Romania. When a tailings dam fails, huge volumes of contaminated water and sediment are rapidly released into river systems (Macklin et al., 1999; Grimalt et al., 1999; Hudson-Edwards et al., 2001) resulting in extensive river pollution, ecosystem damage and loss of life (Diehl, 2001).

The tailings dam failures that occurred in January and March 2000, in Maramureş County northwest Romania, discharged contaminated water and sediment directly into tributaries of the Tisa River, a major tributary of the River Danube. The dam failures resulted in widespread pollution (especially by cyanide) in Hungary, Yugoslavia and Bulgaria and as a result they attracted enormous media attention, and a number of post-spill environmental assessments were undertaken (e.g. UNEP and EU Baia Mare Task Force). However, because of the uncertainty and concern over the long-term effects of heavy metal pollution, a larger scale and more intensive survey of surface water, river sediment and floodplain sediment in the rivers of Maramureş County was carried out in July 2000.

73

W. Leal Filho and I. Butorina (eds.),
Approaches to Handling Environmental Problems in the Mining and Metallurgical Regions, 73–83.
© 2003 *Kluwer Academic Publishers. Printed in the Netherlands.*

The findings of this survey have recently been reported (fluvio, 2000; Macklin et al., in press), and the purpose of this paper is to summarise the principal results and to assess the long-term fate and environmental significance of contaminant metals released by mine tailings dam failures in Maramureş County.

Study Area

Maramureş County has a very long history of base (Cu, Pb, Zn) and precious (Ag, Au) metal mining, exploiting Neogene age hydrothermal vein mineralisation deposits. At present there are 19 flotation plants and 215 disused and functioning mine tailings ponds in addition to a number of Cu and Pb smelters. Since May 1999 cyanide has been used to recover Au and Ag from old tailings material outside the city of Baia Mare and a new tailings pond was constructed near the village of Bozanta Mare to store waste from this new plant (Figure 1). However, high snowfall, followed by heavy rain and rapid snowmelt, caused the tailings pond to fail on 30th January 2000, resulting in the release of 100,000 m³ of waste water and sediment containing high concentrations of cyanide and contaminant metals into the Lăpuş and Someş rivers. Similar weather conditions preceded the second tailings dam failure on March 20th 2000 in the Novat valley, 10 km north of Baia Borsa (Figure 1). On this occasion, 100,000 m³ of contaminated water and 40,000 tons of solid waste were discharged into the Vaser and Vişeu rivers, the latter joining the River Tisa at the Ukrainian border.

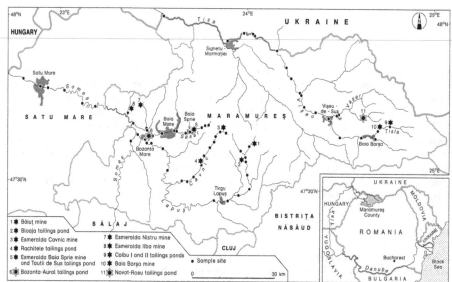

Fig. 1. Maramureş County showing the principal study rivers, towns, tailings dam accident sites and the water / sediment sample site locations.

Mine tailings in both affected river systems were dispersed under very high river flows, during which sediment-associated metals were deposited in both within-channel and overbank environments. The volume of solid waste in relation to the size of both drainage basins was, however, relatively small and as a consequence it did not disrupt river dynamics except immediately downstream of the dam failure sites.

Environmental trends

Maramures County has a complex geological structure, which partially explains the diversity and richness of the mineral ores from this area. The Neogene volcanic rocks are the main characteristic of the area, being found in large quantities especially in the south-western part of the county. They form a well-individualized geological unit, which is generally oriented in an east-west direction, over 100 km in length. The Mesozoic crystalline unit of the Maramures and Rodnei Mountains is found in the north-eastern part of the county, while the central-eastern part belongs to the cross-carpathian flysh. The Maramures and Baia-Mare Depressions consist of Cretaceous and Pannonic molasses deposits.

The main rivers that run through Maramures County are the Lapus, Somes and Viseu Rivers. Lapus, a tributary of the Somes, is the longest river in the county and gathers the waters that run through the Ignis and Tibles volcanic mountains (these mountains consist of lower Paleogene rocks). Along its mountain tributaries there are a series of mining exploitation points. Farther downstream, it crosses the Miocene deposits (marls, limestones, and tuffs).

The Somes River is a tributary of the Tisa River. It runs through the south-western part of Maramures County where it collects a series of small rivers that cross the Neogene eruptive massif of Tibles Mountain, the crystalline massif of Meses Mountain and the Pliocene-Quaternary deposits of the Pannonian Depression.

The Viseu River, another tributary of the Tisa, flows into the Tisa right on the border of the country and its main tributary river is the Vaser. The Viseu, between the emergence and the mouth of the river, crosses first an area of metamorphic rocks (micaschists and paragneiss), then the Paleogene-Oligocene flysh deposits (sandstone, marls and clay and marls rocks) and the Miocene-Pliocene molasses (conglomerates and sands).

The Rivers Lapuş and Someş in the upper Tisa basin drain the actively mined Baia Mare region in Maramureş County. The mining of base and precious metals at Băiuţ in the headwaters of the Lapuş has been previously shown (Fluvio, 2000; Macklin et al., in press) to have degraded the local riverine environment. Downstream transport of mining associated contaminant metals could potentially introduce pollutants into the Hungarian River Tisa (Figure 2).

From the pluviometer point of view, the analysed area, covering most of the Baia-Mare Depression is rich in precipitation throughout most of the year. This is due to the fact that humid air masses, which are part of the Azores Anticyclone, pass over the mentioned area on their way to the eastern part of the continent. The annual average precipitation registered at Baia Mare during 1875 – 2000 period is 941,3 mm. The maximum amounts of precipitation measured at relatively short time intervals (24, 48 and 72 hours) are one of the most important parameters for the activities in the region and were taken into account for the 1961 – 1999 period as well as for January and February, 2000.

Based on those figures, the maximum amounts of precipitation registered in January and February, 2000, when the technological accidents occurred at the two tailing dams, are not the highest among all the observation data, and thus, they do not represent reference values for the analysed data (Dragota et al., 2002).

Field sampling and laboratory metal analysis

In July 2000, 65 surface water, 65 fine grained (<2 mm) river channel and 45 floodplain sediment samples were collected from rivers in Maramureş County affected by industrial, mining and municipal metal pollution (Vişeu, Tisa, Lapuş and Someş), as well as from their principal tributaries (Cavnic, Sasar, Tisla and Vaser) (Figure 1). Water samples were filtered through 0.45 μm filter papers and acidified with three drops of 50% nitric acid in the field before multi-element analysis using an ICP-MS (VG Elemental Plasma Quad II+) was performed. River channel and floodplain sediment samples were collected using a stainless steel trowel from bar surfaces and river banks, respectively. Sediment samples were air dried, sieved through a 2 mm plastic mesh, digested in nitric acid and metal levels determined using AAS (Perkin-Elmer 2380). Channel sediment samples (<63 μm fraction) were subjected to a sequential extraction procedure (SEP) based on a method described by Ure et al. (1993).

Results

Metal concentrations in surface water were assessed against target and imperative values in the EC directive (75/440/EEC) as required of surface water intended to be used for the abstraction of drinking water (Table 1). Metal levels in river and floodplain sediment were compared with the latest (4 February 2000) Dutch target and soil remediation intervention values (Table 2).

The Dutch intervention values for soil / sediment remediation are considered to be numeric manifestations of the concentrations above which there can be said to be a case of serious contamination. These values indicate the concentration levels of metals above which the functionality of the soil for human, plant, and/or animal life may be seriously compromised or impaired. Target values indicate the level at which there is sustainable soil quality and which gives an indication of the benchmark for environmental quality in the long term on the assumption of negligible risk to the ecosystem.

Table 1: EC target and imperative values for the abstraction of surface water for drinking (75/440/EEC).

	Target Value (μg/l)	Imperative value (μg/l)
Pb	—	50
Zn	500	3,000
Cu	20	50
Cd	1	5

Table 2: Target values and soil remediation intervention values for selected metals from the Dutch Ministry of Housing, Spatial Planning and Environment (VROM, 2001).

	Target Value (mg/kg)	Intervention value (mg/kg)
Pb	85	530
Zn	140	720
Cu	36	190
Cd	0.8	12
Values have been expressed as the concentration in a standard soil (10% organic matter, 25% clay).		

Surface water, river and floodplain sediment contamination in Maramureş County

Figure 3 plots as a series of stacked bars, the percentage of surface water, river and floodplain sediment samples in all rivers that fall below (white bars) or above (grey bars) target values, or that exceed (black bars) imperative/intervention values. Although over 50% of surface water samples in Maramureş County have metal concentrations (Pb, Zn, Cu, Cd) that fall below target values, over 20% of samples exceed imperative values for Zn and Cu, and over 30% of samples exceed imperative values for Cd. In contrast to the water samples, much higher percentages of river and floodplain sediment samples in Maramureş County exceed intervention values. Over 50% of river sediment samples and over 40% of floodplain sediment samples exceed Pb and Zn intervention values, and, generally, river sediment samples are more contaminated than floodplain sediment samples.

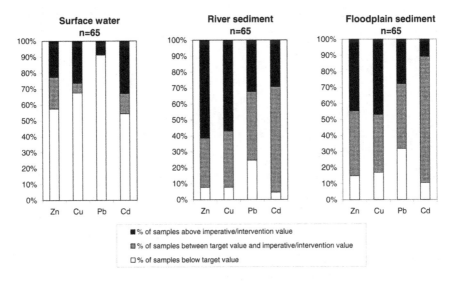

Fig. 3. Percentage of surface water, river channel and floodplain sediment samples that fall below (white bars) or above (grey bars) target values or that exceed (black bars) imperative/intervention values. (n = number of samples).

Pattern of intervention/imperative value excess in Maramureş County rivers

If the samples that exceed imperative/intervention values are examined in more detail, it is apparent that not all rivers in Maramureş County are equally contaminated. Figure 4 plots on a river-by-river basis the percentage of surface water, river and floodplain sediment samples which exceed heavy metal (Pb Zn, Cu, Cd) imperative/intervention values. In the Vişeu/Tisa catchment only sample sites on the Tisla, immediately downstream of Colbu 1 and 2 tailings ponds, exceed imperative Cu and Cd. However, in the Lăpuş/Someş system, the Cavnic, Sasar and Lăpuş all have sites where Zn, Cu and Cd concentrations exceed intervention values. All these rivers have active mines and tailings ponds located on them (Figure 1), and it is interesting to note that even though these rivers are tributaries of the Someş, metal concentrations in all the Someş water samples do not exceed imperative values. Solute contaminants from the Lapuş system are not transported into the River Someş, or across the Romanian-Hungarian border into the River Tisa, which can be termed 'unpolluted' with respect to As, Cd, Cu, Pb or Zn. However, significant solute Hg concentrations have been measured in the River Tisa (Figure 5), the location of peak concentrations suggesting that pollution sources are located within Hungary and are seemingly not related to upstream mining activity.

Zn and Cu concentrations in river sediments exceed intervention values on all rivers in Maramureş County, except the Tisa and Someş. The Tisa was the only river sampled in July 2000 where no river sediment samples exceeded intervention values for Pb, Zn, Cu or Cd. Pb, Zn, Cu and Cd concentrations in floodplain sediments on the Cavnic, and Pb and Cu concentrations on the Sasar, exceed intervention values at over half the sites sampled. It is highly likely, therefore, that floodplain sediment in Maramureş County is a major secondary source of contaminant metals, which may be re-introduced into river channels by land disturbance or by river bank erosion.

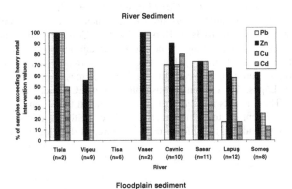

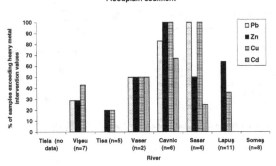

Fig. 4. Percentage of surface water, river channel and floodplain sediment samples that exceed imperative/intervention values on a river-by-river basis (n = number of samples)

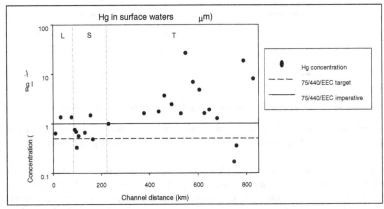

Fig.5: Solute Hg concentrations in the Rivers Lapuş, Someş and Tisa.

Downstream changes in Zn concentration

To pinpoint the exact location of samples that exceed target or imperative/intervention values, Zn concentrations in surface water, river sediment and floodplain sediment have been plotted on a

80

downstream basis for the Vişeu/Tisa and Lăpuş/Someş rivers (Figure 6). Although only Zn data are plotted, the other heavy metals (Pb, Cu, Cd) discussed above exhibit very similar downstream concentration patterns. In the Vişeu/Tisa system, surface water target values are only exceeded at three sites, all on the Tisla close to Baia Borşa mine and Colbu 1 and 2 tailings ponds. The headwaters of the Lăpuş, near to the Băiuţ Mine and Bloaja tailings pond, have Zn concentrations that exceed imperative values (Figure 4), but concentrations rapidly fall to below target values within 10 km of these pollution sources. In the lower Someş, however, downstream of its confluence with the Lăpuş, Zn concentrations fall within EC guidelines for drinking water abstraction purposes. These results indicate that surface water in the Vişeu, Tisa and lower Someş rivers would appear at present not to be of concern to public health, but that surface waters close to active mines and tailings ponds are not safe for use as a source of drinking water.

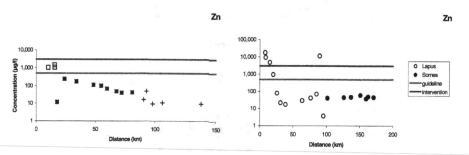

Figure 6. Downstream changes in Zn concentration in water samples in the Viseu/Tisa and Lapus/Somes rivers

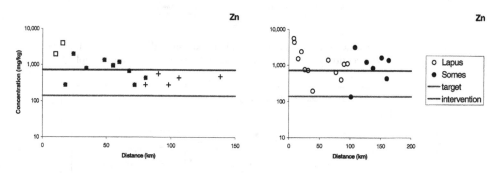

Downstream changes in Zn concentration in river sediment samples in the Viseu/Tisa and Lapus/Somes rivers

Fig. 6: Downstream changes in Zn concentration in surface water, river channel and floodplain sediment samples in the Vişeu / Tisa and Lăpuş / Someş catchments.

In the upper Vişeu catchment Zn concentrations in river sediments are elevated up to 60 km downstream of Baia Borsa mine and Colbu tailings ponds (Figure 6). Zn concentrations in the Vişeu decrease downstream as far as its confluence with the Vaser, where concentrations rise as a result of metal inputs from both municipal wastewater at Vişeu de Sus and the Novat-Rosu tailings pond. In the lower Vişeu and in the Tisa, contemporary river sediment metal concentrations do not exceed intervention values at any of the sample sites. River sediment metal concentrations are generally higher in the Lăpuş / Someş catchment, with the upper reaches of the Lăpuş near to Băiuţ Mine and Bloaja tailings pond particularly contaminated (Figure 6). Although Zn concentrations decline downstream of these pollution sources, they rise again downstream of both the Cavnic and Sasar tributaries and exceed the Zn intervention value on the Someş at the Hungarian border.

Metal and As concentrations in within-channel river sediments in the River Lapuş exhibit a similar pattern to surface waters, with peak concentrations occurring immediately downstream of the Băiuţ mine, and then decreasing downstream (Figure 7). This suggests sediment associated contaminants are not being greatly mobilized down the River Lapuş. The River Sasar provides a secondary source of pollution in the lower Lapuş (Figure 7). In the River Someş, Cu and Zn are present in concentrations greater than Dutch imperative guidelines, with As, Cd and Pb falling between target and imperative guideline values. Whilst sediment-bound metal pollutants are not being transported from Maramureş County into the Tisa at contaminant levels, concentrations are greatest in the Tisa immediately downstream of the Someş – Tisa confluence. Cd (Figure 7) is representative of Cu and Zn in that concentrations fall between target and imperative guidelines.

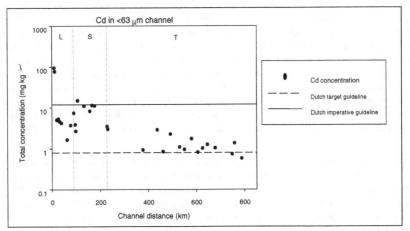

Fig. 7. Concentrations of Cd in river channel sediments in the Rivers Lapuş, Someş and Tisa

Speciation of channel sediment-bound metals

Cd and Zn were found to be weakly adsorbed to sediments in the 'exchangeable' phase, and hence can be seen as being more readily bioavailable and potentially toxic than a more strongly adsorbed element such as As. Figure 8 indicates the sites particularly threatened by Cd pollution, a majority of the Lapuş - Someş - Tisa reach can be considered to be moderately affected. For As, Cu, Pb and Zn 'bioavailable' concentrations are not significant, unlike in the Rivers Lapuş and Someş. 'Bioavailable hotspots' are present in the Lapuş for Cd, Cu, Pb and Zn, related to metal mining activity at Băiuţ. Cu 'bioavailable hotspots' are also present at four sites in the Someş, although it is unlikely that the Băiuţ mine is the lone source of this pollution problem.

Floodplain sediment metal concentrations in the Vişeu / Tisa and Lăpuş / Someş catchments are generally lower than those found in contemporary river sediments (Figure 6). In the Lăpuş and Someş catchments, floodplain and river sediment metal concentrations generally parallel each other, although floodplain metal levels decrease more rapidly downstream and at none of the Someş sites do metal concentrations exceed intervention values. Nevertheless, in the upper Lăpuş (and the Cavnic and Sasar), floodplain sediments do appear to be significantly contaminated. There is a very distinctive relationship between metal concentrations in contemporary river and floodplain sediment in the Vişeu and Tisa catchments. Present day river sediment metal values progressively decrease from 60 km downstream and remain generally low to the Ukrainian border. Floodplain metal concentrations, however, increase at this point and progressively rise in the lower part of the Vişeu valley until declining downstream of the confluence with the River Tisa (river km 80). It is not clear what the source of the pollution is, but it certainly pre-dates the March 2000 Novat-Rosu mine tailings pond spill.

Conclusions

In July 2000 surface water metal levels in the Vişeu, Tisa (within Romania), and Someş (downstream of the Lăpuş confluence) rivers complied with the EC water quality directive. The upper parts of the Tisla, Lăpuş and Cavnic basins, and the middle reach of the Sasar, however, are severely polluted, mainly as the result of present and past mining operations. Metal concentrations in surface water at some sites within all of these catchments represent a potentially serious hazard to human health. River sediment metal concentrations at over half of the sites sampled in the Vişeu / Tisa and Lăpuş / Someş catchments exceed Dutch intervention values for soil remediation. Cu and Zn concentrations are particularly high in the Cavnic, upper Lăpuş, Sasar, and Tisla catchments. Elevated levels of both Cu and Zn are also found in the River Someş (down to the Hungarian border) and in the lower Vişeu, indicating that sediment-associated metals are currently being dispersed away from the main mining areas. Although floodplain metal concentrations are generally lower than those found in river sediment in the Vişeu / Tisa and Lăpuş / Someş catchments, metal concentrations do exceed intervention values in many samples, with Cu and Zn appearing to pose the greatest hazard especially in the mining-affected Cavnic, upper Lăpuş and Sasar valleys.

The long-term environmental impact of the Bozanta-Aural and Novat-Rosu mine tailings pond failures is still difficult to quantify, primarily because of the multiple sources (past and present) of contaminant metals affecting the region's rivers. Although metal concentrations in surface water, river sediment and floodplain sediment are significantly elevated immediately downstream of the spill sites, within less than 10 km metal values fall appreciably. Indeed, contamination from present mining activity in the Cavnic, upper Lăpuş, Sasar and Tisla catchments is probably having a more significant impact on water and sediment quality than the tailings dam failures themselves. Whilst not underestimating the short-term ecological impacts of the spills, they should be seen in the context of adding to much longer-term contamination arising from decades of poorly regulated, and largely untreated, industrial, mining and urban discharges into local rivers.

References

Diehl, P. (2001) World Information Service on Energy (WISE) Uranium project [online]. Available from: http://www.antenna.nl/wise/uranium/ [Accessed 20 February 2002].

Dragotă, C., Bălteanu, D. (2002) *Regimul precipitaţiilor atmosferice şi hazardele pluviometrice în Depresiunea Baia Mare*, Revista Geografică, t. VIII, 2002, pp. 25-32

Fluvio: *The long-term fate and environmental significance of contaminant metals arising from the January and March 2000 mining accidents in Maramureş County, Romania*, 2000. Unpublished report prepared for the EU Baia Mare Task Force, 38pp.

Grimalt, J.O., Ferrer, M. and Macpherson, E. (1999) *The mine tailing accident in Aznalcóllar*. In: *Science of the Total Environment*, 242, 1999, pp. 3-11.

Hudson-Edwards, K.A., Macklin, M.G., Miller, J.R. and Lecher, P.J.*(2001) Sources, distribution and storage of heavy metals in the Rio Pilcomayo, Bolivia.* In: *Journal of Geochemical Exploration,* 72, 2001, pp. 229-250.

Macklin, M.G., Hudson-Edwards, K.A., Jamieson, H.E., Brewer, P.A., Coulthard, T.J., Howard, A.J., and Remenda, V.H. (1999) *Physical stability and rehabilitation of sustainable aquatic and riparian ecosystems in the Rio Guadiamar, Spain, following the Aznalcóllar mine tailings dam failure.* In: Rubio, R.F. (Ed.) Mine, Water and Environment, International Mine Water Association, 1999, pp. 271-278.

Macklin, M. G., Brewer, P.A., Balteanu, D., Coulthard, T.J., Driga, B., Howard, A.J. and Zaharia, S. (in press): *The long term fate and environmental significance of contaminant metals released by the January and March 2000 mining tailings dam failures in Maramures County, upper Tisa Basin, Romania.*

Ure, A.M., Quevauviller, P., Muntau, H. and Griepink, B. (1993) *Speciation of heavy metals in soils and sediments,* An account of the improvement and harmonization of extraction techniques undertaken under the auspices of the BCR of the Commission of the European Communities. In: International Journal of Analytical Chemistry 51, 1993, pp. 135-151

VROM, Intervention and target values – soil quality standards [online], 2001. Available from: http://www.minvrom.nl/minvrom/docs/bodem/annexS&I2000.PDF. [Accessed 28 June 2001].

CHAPTER 11

Environmental Protection in the Steel Industry, exemplified by ThyssenKrupp Stahl AG, Duisburg

Dr. Gunnar Still

Corporate Division Environment
ThyssenKrupp Steel AG
Kaiser-Wilhelm-Str. 100
47166 Duisburg
Germany

Summary

ThyssenKrupp Steel ranks among the leading steel producers. The prominent position of the company, also in the field of environmental protection, is generally acknowledged among experts. The reasons for ThyssenKrupp Steel's special advocacy of environmental protection lie in the location of the parent works in the densely populated Rhine and Ruhr region, the direct proximity of residential areas, and the voluntary commitment of the company to improve the quality of life of the local residents and its employees through air pollution control and noise abatement measures. The housing density has additionally been an important criterion for the development of waste management structures, thereby intensifying the pressure to recycle residual materials and substances and also opening up, on the other hand, possibilities of using these materials and substances in the building and construction industry. New fields of technological environmental protection world-wide include climate protection and, in the European Community and Germany, for example, soil conservation. Technological environmental protection is being accompanied increasingly by organisational measures, of which environmental management is representative. The main means of air pollution control are described, the importance of recycling for steel making explained, and product- and production-specific environmental protection discussed.

Introduction

The integrated iron and steel mill operated by ThyssenKrupp Stahl at its site in Duisburg is the second largest in the world after that of the POSCO Group in Kwangyang. The company's core activity is the production of flat steel. The mix of finished steel products ranges from heavy plate to sheet with highly sophisticated coatings. ThyssenKrupp Stahl also includes enterprises that are active in downstream customer-driven manufacturing stages [Kohler, 2000].

ThyssenKrupp Stahl aims to concentrate the company's entire liquid metal stage in Duisburg. It is here that the steel making activities, from coal and ores through to the hot strip and downstream processing operations, have been consolidated. The most important technical statistics of the plants and facilities operating in Duisburg and making primary and intermediate products are listed in **Table 1, Figure 1,** [Philipp, 2000a]. In the fiscal year 2000/2001, three sinter plants with an annual output of 11.2 mill. t sinter, 4 blast furnaces with an annual hot metal production of 10.7 mill. t, and two basic oxygen furnace (BOF) steel plants equipped with two 380 t and three 265 t converters, respectively, were in operation in Duisburg. The rolling reduction work takes place at two hot strip mills and follow-on cold rolling mills. The focal points of ThyssenKrupp Stahl's technological efforts are metal-protective coating by means of hot-dip, electrolytic and organic processes, and

85

W. Leal Filho and I. Butorina (eds.),
Approaches to Handling Environmental Problems in the Mining and Metallurgical Regions, 85–106.
© 2003 *Kluwer Academic Publishers. Printed in the Netherlands.*

applied technologies as used in the production of tailored blanks, composite materials and hydroformed components. The downstream processing into various end products is distributed among several sites that lie within a maximum radius of 100 km of the iron and steel mill.

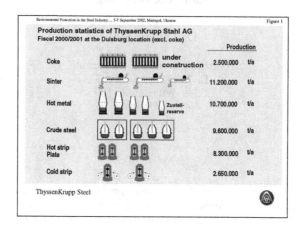

Table 1. Metallurgical plants at Thyssen Krupp Stahl AG's site in Duisburg in fiscal year 2000/2001

Coking plant				
Designation/parameter	Dimension	Battery 2	Battery 6a	Battery 6b
Year of construction		1983	1971	1974
Oven chambers	Number x chamber charge [m³]	60 x 15	52 x 28.5	52 x 28.5
Output	[t/a]	1,300,000		

Sinter plant				
Designation/parameter	Dimension	Schwelgern 2	Schwelgern 3	Schwelgern 4
Year of construction		1964	1970	1979
Strands	Number x suction areas [m²]	1 x 150	1 x 444	1 x 250
Output	[t/a]	2,000,000	5,900,000	3,300,000

Blast furnace plant					
Designation/parameter	Dimension	Hamborn 4	Hamborn 9	Schwelgern 1	Schwelgern 2
Year of construction		1964	1962	1973	1993
Hearth diameter	[m]	10.7	10.2	13.6	14.9
Working volume	[m³]	2030	1833	3844	4769
Output	[t/a]	1,700,000	1,700,000	3,400,000	3,900,000

BOF steel plants			
Designation/parameter	Dimension	Bruckhausen	Beeckerwerth
Year of construction		1969/98	1962/70
Converters	Number x heat weight [t]	2 x 380	3 x 265
Output		4,600,000	5,000,000

Rolling reduction and downstream process lines	
Casting-rolling plant	Cold strip finishing shops
Hot strip mills, lines for hot strip processing by means of pickling, skin-passing, slitting and edge dressing	Lines for hot-dip, electrolytic and organic coating
Cold strip mills	Plate mill
Batch and continuous annealing furnaces	

Any description of ThyssenKrupp Stahl AG's metallurgical plant structure should include a mention of these facilities' special characteristics. The group of blast furnaces in Schwelgern also includes Schwelgern No. 2 blast furnace, which was built in 1993 and, with a hearth diameter of 14.9 m and a working volume of 4,769 m^3, is the world's third-largest blast furnace, [Schulz,1995]. The latest important step in the company's technological development was the start-up, in April 1999, of a casting-rolling plant (CSP) after a two-year construction period [Hendricks, 2000a]. ThyssenKrupp Stahl AG thereby became the first company to install a casting-rolling plant in a conventional integrated iron and steel mill. The production from the BOF steel plant operating the two 380 t converters is processed in the CSP in a combined production stage, **Figure 2**.

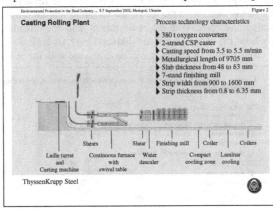

The company has taken the decision to replace the now ageing AugustThyssen coking plant by a new facility, [Hendricks, 2000b]. The new coking plant in Schwelgern will, with two batteries, realize the same output of 2.5 mill. t blast furnace coke as the six batteries still needed in the old coking plant. The most important data are compiled in **Table 2**. With a useful chamber volume of 93 m^3, the plant will have the largest coking chambers ever built.

Table 2   **Construction and operating data of the coke-oven batteries at the coking plant in Schwelgern**

	Numerical value	Dimension
Number of batteries	2	[1]
Number of chambers	70	[1]
Chamber height	8.43	[m]
Chamber width	0.59	[m]
Chamber length	20.9	[m]
Useful chamber volume	93	[m^3]
Coal charge per chamber (moist)	78	[t]
Average coking time	25	[h]

88

Today, almost 50 % of the liquid steel produced in the Federal Republic of Germany comes from the steel city of Duisburg. A significant feature of the operating location is the close co-operation between ThyssenKrupp Stahl and Hüttenwerk Krupp Mannesmann GmbH, in which Thyssen Krupp Stahl holds a participating interest. The two companies form a state-of-the-art production and plant configuration which, through its location, results in favourable transport costs. This location offers optimum conditions for the sourcing of raw materials such as ores and coal via the port of Rotterdam, as well as for steel exports to overseas via the port of Antwerp. An important location factor is the proximity to qualified steel fabricators in North Rhine-Westphalia as well as to European and German customers [Kohler, 2000].

TECHNOLOGICAL ENVIRONMENTAL PROTECTION

The topic of the present report is the protection and conservation of the environment. It goes without saying that the close spatial integration of the production facilities into neighbouring residential areas, due to the fact that Duisburg is a densely populated conurbation, presents particular challenges here. This is especially evident in the shown photo, above all when it is considered that the prevailing wind direction is south-west, and the residential development to the north-east of the works site is directly in the lee of the metallurgical facilities, **Figure 3**. The main task involved in protecting and conserving the environment, and one already identified at an early stage, has consequently been the reduction of dust-like and gaseous emissions at the Duisburg site. The measures taken or envisaged in this respect will be discussed with the aid of examples.

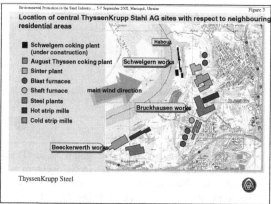

EMISSION REDUCTION MEASURES

Fume and dust control in steel production. The development of dedusting technology was initially closely associated with the way that steel making developed. A decisive breakthrough in the reduction of dust emissions in this field came with the introduction of basic oxygen furnace steel making and the accompanying decline of the basic-Bessemer and open-hearth processes. The last basic Bessemer heat in the Federal Republic of Germany was tapped in 1966, and tapping of the last open-hearth heat in West Germany took place in 1982.

The first basic-oxygen converters came ready-equipped with effective systems for extracting the dust from the primary off-gas. This step was made possible by the considerably decreased volume of off-gas generated in this process. The latest step in the development of fume and dust control at basic oxygen converters has been dry-type dedusting combined with converter off-gas recovery,

Figure 4, [Philipp,1987a], [Schulz, 1992], [Philipp, 2000b], [Philipp, 2001a]. The converter off-gas is dedusted by dry-type electrostatic precipitators to a dust level of < 10 mg/m³, after which the off-gas, with a yield of around 0.72 GJ/t crude steel, is used in the energy grid of the Duisburg iron and steel mill. Further advantages of the process lie in the almost waste-water-free operation of the system and thus in the avoidance of sludge from wet-type waste gas cleaning, and in the re-utilisation of the accumulating dust.

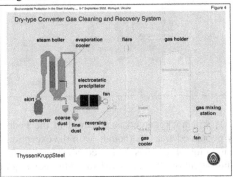

Secondary dust removal. Since the 1970s, primary dedusting has been complemented increasingly by the removal of so-called diffuse dust sources through the use of secondary dedusting systems. Besides the dedusting of the steps involved in the actual steel making process, such as the reladling pits at the hot metal production stage and secondary metallurgy facilities at the steel making stage, mention should be made at this point particularly of the internal dedusting of sinter plants, the extraction of dust from stock- and casthouses at blast furnaces and, last but not least, of the removal of dust from slab scarfing machines, [Philipp,1987b]. The humidifying of ore and coal stockyards has also made an important contribution towards reducing the dust emissions.

Results of dedusting and trends of development. The diverse dedusting measures, above all in the natural resources industry, have led to a major reduction in dust emissions and improvement in air quality and, especially, to a decrease in dust deposits and airborne dust concentrations. While the level of production has remained almost unchanged, the dust emissions at the location of Duisburg, for example, have fallen from 5 kg/t crude steel in 1975 to less than 0.5 kg/t crude steel today, Figure 5. The slight rise in dust emissions from 1997 is attributable to the closure of the steel plant at the Dortmund works and the corresponding increase of production in Duisburg. Measures designed to improve the air quality further have meanwhile been initiated here at an investment of 60 mill. €. Ambient airborne dust concentrations also exhibit an asymptotic trend in Duisburg, as do the depicted dust emissions. Further dust reductions will consequently be achievable only at high technical and economic expense which will yield only little improvement in the ambient air quality situation.

90

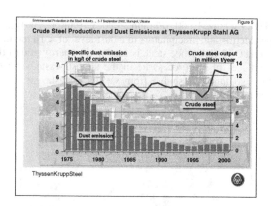

The dust issue, however, is still not regarded as resolved in the Federal Republic of Germany, particularly in the densely populated areas. Although, with the exception of a few locations near to emission sources, the precipitation of dust has decreased considerably in importance, fine respirable dust is attracting increasing attention. This has led to additional requirements at EU level and, especially, with regard to national implementation. The air quality and emission standards (TA Luft) that have just been adopted, for example, lower the annual average for airborne particulate matter concentrations from, previously, 150 µg /m^3 to, in future, 40 µg/m^3, measured in terms of respirable dust (PM$_{10}$), [TA Luft.2002] At the same time, the daily average is to be restricted, allowing 35 transgressions of a value of 50 µg/m^3 annually. A further reduction to 20 µg/m^3 is being targeted, (Ri 1996 L 61).

Process developments aimed at the reduction of emissions

Emissions reduction technology has meanwhile reached an advanced state. Nevertheless, gaps become evident, making further research and development work necessary. A few projects aimed at the reduction of

- dust emissions
- emissions of carcinogenic substances, and
- organic trace gases

will be discussed in brief. A common feature of these projects is the close association with the production processes, such that these developments can be termed as production-integrated environmental protection.

Coke Stabilisation Quenching. The coke dry quenching process applied today in many coking plants, and preferred by Russia and Japan, was developed in the former USSR. The reasons for this lay essentially in the operational difficulties involved in the wet quenching of coke under frosty conditions. The use of the process in Japan and East Asia, by comparison, is governed primarily by energy management considerations, as coke dry quenching makes it possible to utilize the waste heat, although at high plant-related expense.

The decision to install a wet quenching system at the new Schwelgern coking plant was influenced by the considerable progress made in wet quenching, the good operating experience gained in the simultaneous use of top and bottom quenching at Hüttenwerke Krupp Mannesmann GmbH, the great technical effort and expense involved in coke dry quenching, the still inadequate availability of coke dry quenching facilities from the viewpoint of a modern iron and steel mill, the high charge weight of the new coking plant, and other company-related reasons. These considerations induced ThyssenKrupp Stahl to develop the combined top and bottom quenching process further with specialist enterprises and to adopt it, under the name of Coke Stabilisation Quenching (CSQ process), for the newly constructed Schwelgern coking plant.[Hendricks, 2000b].

In the CSQ process roughly one-third of the quenching water is applied onto the hot coke via nozzles arranged over the quenching car. The water for bottom quenching is supplied from quenching water tanks, via downpipes and a spacious water header with outlets on the bottom of the quenching car, into the coke charge, **Figure 6**. This method of water quenching cools the coke down much faster and leads to a substantial decrease in the emissions of H_2S and CO as compared with classic wet quenching, [Hein.2000], [Toll.2000], [Nelles,2000.a], [Nelles.2000.b], [Nelles. 2001].

The decisive advantage of the CSQ process, however, is deemed to be that the specific emission of dust is reduced further by a considerable degree compared with the wet quenching methods used so far. The dust is removed via two additional scrubbing stages, each comprising downstream lamella-type separators integrated in the quenching tower, making it possible to reduce the residual dust content from, previously, 50 g/t_{coke} and, for progressive wet-quenching processes, from 25 g/t_{coke}, to, in the meantime, residual dust contents of less than 10 g/t_{coke}, [Still, 2002]. There is, however, no significant difference in the dust emissions from CSQ wet quenching and coke dry quenching. Of relevance in terms of environmental health is that the dust emitted during coke dry quenching is a fine, respirable dust with a grain diameter of less than 10 µm, while, according to the studies available so far, the dust emitted during wet quenching has a coarser grain distribution, (Masek.1975). Further investigations in this area are planned after the start-up of the CSQ wet quenching system.

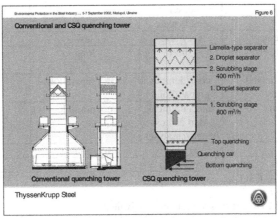

The advantages and disadvantages of coke dry quenching and of CSQ coke wet quenching are compared in **Table 3,** and the emissions in **Figure 7**, [Philipp. 2001.b].

92

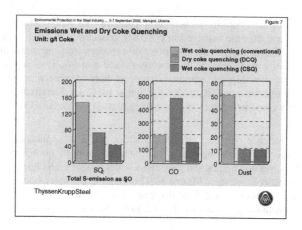

Table 3:   **Comparison of coke dry quenching and CSQ coke wet quenching**

Criterion	Coke dry quenching	Coke wet quenching CSQ process	CSQ process in comparison with coke dry quenching [1]
Emissions [2]			
Total dust	$19.9^{2\,u.\,3)}$	$18.6^{2\,u.\,3)}$	=
Fine respirable dust <10 μm	$5.5^{3\,u.4)}$	$<1^{3\,u.4)}$	+
SO_2	$72^{2\,u.\,3)}$	$40^{2\,u.\,3)}$	
CO	$480^{3u.\,4)}$	$150^{3\,u.\,4)}$	+
Plant safety	special structural and operational measures	no special requirements	+
Plant availability	inadequate	high	+
Maintenance requirement	very high	low	+
Service duration between repairs	short	long	+
Allowance for start-up and shut-down procedures	considerable preparations	short preparations	+
Waste heat utilisation	uneconomical based on cost-benefit analyses	not possible	no evaluation
Capital outlay	very high	reasonable	+
Operating costs	very high	reasonable	+

[1]   Symbols mean: + CSQ process is better
 - CSQ process is worse
 = CSQ process and CDQ process are roughly equivalent
[2] The quantitative emission data relate to the selected limits. These range from the transfer of the hot coke after pushing, through to and including the coke treatment. Data as per DMT expert opinion dated 6.03.2001.[17])
[3] Values in g/t_{coke}
[4] Value relates to the quenching stage only.

Economic efficiency is of paramount importance where investment decisions are concerned. The comparison of the two quenching processes showed that coke dry quenching would have required a capital outlay of 150 mill. € for the envisaged coking plant, whereas the capital expenditure for the CSQ coke wet quenching process is 50 mill. €. The calculation of the operating costs led to a similar result, after allowing for the additional cost of coke dry quenching at 7.50 €/t$_{coke}$. Included here is the amount credited for possibly generated electric current, which would, however, have to be fed into an external grid system on account of the company's surplus situation.

The findings of the company's own economic assessment have been confirmed by the Commission of the European Communities in its documentation regarding the application of best available techniques for protecting the environment. According to this, coke dry quenching plants cannot be operated economically in the European Union [EIPPC Bureau.2000].

A further and, as yet, insufficiently considered advantage of coke wet quenching is undoubtedly that wet-quenched metallurgical coke has qualitatively superior properties to those of dry-quenched coke. Although the sensible heat of the incandescent coke cannot be used in wet quenching, this disadvantage is more than offset by the coke's superior properties in the blast furnace process.

Single-chamber pressure control at coke oven doors. The emission of carcinogenic substances is receiving increasing attention in all sectors of industry. In this connection, measurements performed in the proximity of coking plants have led to the task of reducing benzol and PAH and, in particular, coke oven emissions. The emission levels of coke ovens are influenced mainly by leaks at the oven doors and at the charging holes in the oven roof. This is essentially a consequence of the changes in the pressure conditions that come about during the coking in the ovens.

ThyssenKrupp Stahl undertook a research project with the objective of keeping the oven chamber pressure constant, irrespective of the coking time and, therefore, of the generation of gas. This was achieved by means of a single-chamber pressure control system, **Figure 8**. The new PROven method makes it possible to control the gas pressure of a freshly charged coke chamber so that, even when the generation of gas is at its greatest, a pressure slightly above atmospheric prevails in the lower region of the doors within the coking chamber, and the leaks via door seals and charging hole covers are appreciably reduced, **Figure 9** [Philipp,1999a], [Hofherr,2001a], [Hofherr, 2001b]. Through measurements at an operational coking plant where the system was installed for trial purposes, an independent specialist institute demonstrated a decrease of over 60 % in door emissions. The new coking plant in Schwelgern will be fitted with the single-chamber pressure control system.

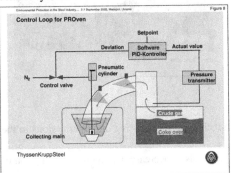

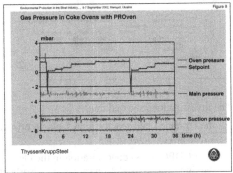

Reduction of sinter plant dioxin emissions. A study jointly undertaken by the Federal Environmental Agency and the German Iron and Steel Institute (VDEh) during the years from 1992 to 1995 revealed that, in 1993, sinter plants accounted for around 95 % of the dioxin emissions generated in iron and steel making in the Federal Republic of Germany, [Batz, 1996], [Theobald,1995]. In the EU, sinter plants are estimated to account for around 18 % of dioxin emissions, [Bröker, 1997].

A working group comprising European steel makers set itself the aim of developing a process to reduce these dioxin emissions. A process with essentially the following characteristics was striven for:

– Possibility of integration in existing waste-gas cleaning systems
– Low floor space requirement with regard to retrofitting at existing plants
– Use of a dry waste-gas cleaning process
– Re-circulation of the removed dust to the sinter strand
– Availability of the sinter plant in the materials cycle of integrated iron and steel mills.

Following preliminary studies, [Kersting,1997], a demonstration plant was built at the 150-m² #2 sinter strand of Thyssen Krupp Stahl AG in Duisburg, **Figure 10**. The sinter strand has a daily output of 6,000 t and a waste gas volume flow averaging 360,000 m³/h. The demonstration plant comprised

– an adsorption stage in the form of an entrained-bed reactor,
– an electrostatic precipitator, and
– a downstream oxidative catalytic converter, **Figure 11**.

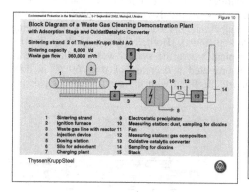

The investigations showed that

– the entrained bed reactor represents an effective technique for reducing dioxin emissions in the waste gas of sinter plants,
– the PCDF/D mass concentration in the cleaned gas tends to decrease with increasing addition of the lignite coke used primarily as an adsorbent,

- the mean separation rate is around 70 %, with PCDF/D mass concentrations between 0.13 and 0.76 ng/m³ I-TEQ/m³ being achieved with higher dosages of lignite coke, and
- only a small percentage of the PCDF/D pollution load was made up of the particularly toxic 2, 3, 7, 8-TCDD, of which in turn 76 % was retained, **Figure 12**, [Philipp, 1999b], [Philipp,1999c], [Philipp,2000c].

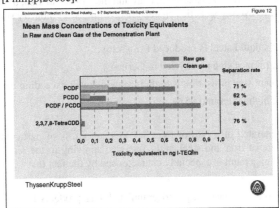

It was not possible to achieve the target value of 0.1 ng/I-TEQ/m³. Sinter strands 1 and 3 are being retrofitted with an appropriate system. The German steel industry expects that, after all the sinter plants have been retrofitted, it will be possible to reduce the current mass flow of PCDF/D emissions to less than 40 g I-TEQ/a.

The downstream catalytic converter failed to fulfil expectations. After the performance of the trials, the separation efficiency deteriorated within a short space of time from 60 %, when new, to the point of ineffectiveness. Reactivation of the catalytic converter was not possible for technical as well as for environmental reasons so that, based on the present state of knowledge, the use of catalytic converters for the reduction of dioxin emissions from sinter plants does not present any prospect of success.

Recycling

Recycling is a key environmental conservation strategy of the iron and steel industry, [Philipp,1992], [Bredehöft, 2001]. Steel making is a process of concentration which, from the preparation of the raw materials through to every stage of production, requires the availability of large quantities of basic and indirect materials. The impact of the steel industry on the environment is therefore, first and foremost, a quantity-related issue, with recycling being the strategy for coping with large flows of materials and their accompanying impacts on the environment in an ecologically acceptable way.

Classic examples worthy of mention include:

- Re-utilisation of scrap
- Use of slag and other residual products
- Re-utilisation of ferriferous residuals
- Re-circulation and multi-stage (cascade) use of water

96

- Reprocessing of pickling fluids and repeated use of rinsing water
- Use of process-generated gases (co-products)

Solid residuals and slag management. The recycling of solid residuals from the iron and steel making processes has reached an advanced state around the world and also, therefore, in the Federal Republic of Germany. Recycling statistics characteristic of the steel industry show that:

- 45 % of the world's liquid steel is produced from scrap,
- 100 % of the blast furnace slag is recycled in Germany,
- the level of steel plant slag utilisation is high, meanwhile well exceeding the 90 % mark, and
- recycling rates are high, at around 95 %.

Metallurgical slag, generated at a rate of over 300 kg/$t_{crude\ steel}$, is of particular importance to the circulatory management of iron and steel mills, this also applying to the recycling rate. The more recent trend in slag management is especially characterised by the fact that

- the processing of blast furnace slag into granulated slag products has increased further by a significant degree,
- efforts continue to be geared to increasing the use of steel plant slag in the building and construction industry, **Figure 13**.

The utilisation of slag in road building and cement production means a corresponding saving in natural raw materials, and the use of granulated sand in the cement industry, for example, leads to a considerably lower energy consumption and reduction in CO_2 emissions, [Philipp.1992].

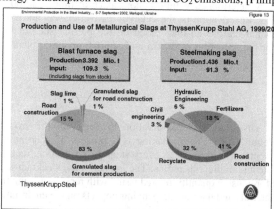

In-plant water management and water conservation. The specific water intake is an important statistic of any resource-saving in-plant water management system geared to the protection and conservation of water. Thyssen Krupp Stahl AG has reduced its water intake at its site in Duisburg to 2.6 m^3/$t_{crude\ steel}$, **Figure 14.** The repeated or multiple use of water conserves water resources. The resulting decrease in the complexity of the waste water treatment and in the waste water load are important for the prevention of water pollution.

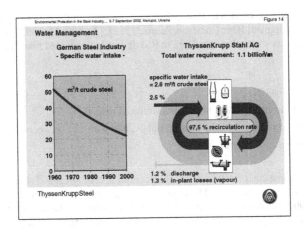

Figure 14

Re-utilisation of ferriferous residuals. As yet, there is no satisfactory solution available, technically, economically or energy-wise, for re-utilising the dust and sludge generated in iron and steel production. Another problem is that it is hardly possible to influence the amount of dust and sludge that accumulates, and that there is no satisfactory guaranteed disposal. A new project is therefore underway with the aim of using shaft furnaces built on the works premises of ThyssenKrupp Stahl to process the roughly 400,000 t dust and sludge accumulating in steel production in the Duisburg region, Figure 15, [Philipp, 2001a].

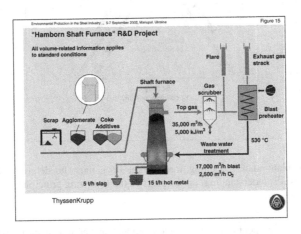

Figure 15

So far, agglomerates consisting of ferriferous dust and sludge pressed together to approximately the shape of sandy limestone have been provided as charge materials.

The operating results achieved to date at a test facility have been so successful that ThyssenKrupp Stahl now intends to build a commercial-scale plant.

Institutional environmental protection

Over the past few years, environmental protection has developed increasingly as a result of new trendsetting approaches. The international community has presented its ideas in this respect in, among others, the declaration of the Brundtland Commission in 1987 and Agenda 21, adopted at the

98

global conference in Rio de Janeiro in 1992 and has called upon all the nations of the world to strive for Sustainable Development, [Our Common Future, 1987], [Agenda 21, 1992]. The world's largest conference in the history of mankind, in 1992 in Rio de Janeiro, regarded mankind's most important tasks to be:

− the fight against poverty,
− the protection of the environment and conservation of resources,
− co-operation between the states, the general public, and with and between social groups, and
− education and scientific research.

It has meanwhile become customary to depict the economic, ecological and social aspects addressed within the context of Sustainable Development in graphic form, **Figure 16**. The task is consequently to optimise the depicted principles and bring them into harmony with one another. Of significance, as far as the topic under consideration here is concerned, is that Agenda 21 from our viewpoint also reveals the limits of environmental protection, which should be integrated into rather than isolated from discussions of other areas of policy [Keating,1993].

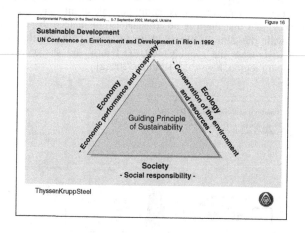

The location of any plant managed along free-market lines must be determined, according to ThyssenKrupp Stahl AG's understanding, on the basis of the following principles within the context of Sustainable Development:

− Preservation and development of the plant's technological and economic performance
− Responsible conduct towards employees, neighbouring areas, shareholders and generations to come, and
− Implementation of a forward-looking policy of preventative environmental protection.

ThyssenKrupp Steel is committed to these principles, which are expressed in ThyssenKrupp Stahl's environmental declaration, **Table 4**.

Table 4 Environmental principles of ThyssenKrupp AG

We see environmental protection as an important corporate objective.
We preserve natural resources.
We use ecologically acceptable production facilities and methods.
We bear responsibility for our products.
We develop environmental protection jointly with our customers.
We research new techniques.
We take part in joint initiatives.
We keep the public informed.
We regard environmental protection as everybody's responsibility.

The development of environmental controls, which is being influenced to a greater extent by political and also academic circles, has led in the past few years to new approaches which are identifiable particularly in an increasingly integrative attitude towards environmental protection. This transformation is being promoted essentially by the directives issued by the European Union. It can be seen that organisational environmental protection has thereby become more defined as well as being an independent component alongside classic technological environmental protection, **Figure 17**.

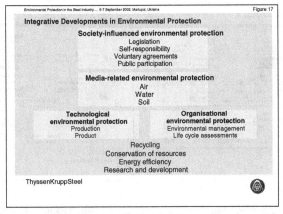

Among the organisational instruments available for protecting the environment we see, in particular, environmental management which, in turn, can include such elements as eco-balances (life cycle assessments), environmental monitoring, and further education.

Environmental management

The successes of the steel makers, and the progress made in ecological conservation, demonstrate that steel makers have long had effective means of environmental management at their disposal. Steel makers around the world are having their environmental management systems certified to ISO standard 14001, [ISO 14001, 1996]. The objective in this respect is to use and further develop the positive elements of modern management in the day-to-day practice of protecting the environment.

Monitoring. An essential part of a good environmental management system is to monitor emissions as well as the main environmental impacts and activities. The tasks involved in such monitoring include:

100

- Performance of emission measurements
- Waste water analyses
- Drafting of waste reports and waste management concepts
- Annual reports by the officers responsible for ambient air protection, water management and recycling
- Emissions declarations

The measurements and reports are in some instances required by law or stipulated in plant permits, or take place at the discretion of the operators or in co-operation with the authorities.

The monitoring of airborne emissions has proven to be especially important and effective, [Philipp,1997]. The most important sources of emissions at ThyssenKrupp Stahl are shown in **Figure 18.** The emission measurement data are relayed to a central computer, which:

archives the measurement data,

enables the retrieval and output of the data,

prints out reports in cases where emission limits are exceeded, and

has the function of transmitting the measurement data online to the remote regulatory bodies, **Figure 19.**

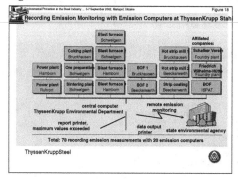

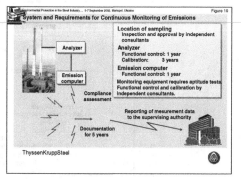

Eco-balances (life cycle assessments). Steel products and materials must also increasingly measure up to environmental standards, [Philipp,1994a], [Philipp,1994b], [Philipp,1996]. Eco-balances/life cycle assessments have shown that the environmental relevance of the use of steel derives

- in case of non-static goods, such as vehicles, mainly from the utilisation of the products from which they are made, and
- in case of static goods, e.g. in the building and construction industry, from the manufacture of the material.
-

ThyssenKrupp Stahl endeavours to provide its customers with such environmental data about its products as are needed for the users' planning and further processing purposes.

Further education. A positive and frequently underestimated element is the sensitisation and further education of employees and of the managerial staff not directly involved in environmental matters. Responsible environmental management must not become exhausted as a result of short-term decisions, but be geared to Sustainable Development and the long-term securement of corporate viability. Environmental protection should be understood, and practised, as a joint responsibility by all levels of management and by every employee.

Research and development

Research and development can be seen as having a role to play in both technological and organisational environmental protection. An additional point which should be regarded as an important area of environmental research and development within this context is the further development of production-specific and product-related environmental protection, [Drewes,1999], [Lindenberg, 1999]. Examples of production-specific environmental protection include the studies described above on the reduction of dioxin emissions and the use of shaft furnaces as a means of recycling.

A focal point of the applied research activities pursued at ThyssenKrupp Stahl is the preliminary stage of automobile manufacture, [Stümpke, 1998]. The joint "Ultra Light Steel Autobody (ULSAB)" project conducted in the past few years with the aim of reducing automobile weight has been a good example of product-related environmental protection. The demonstration auto body developed by 35 producers from 18 countries has, compared with vehicles of conventional design,

- a 25 % lighter weight in relation to similar auto bodies, and
- 80 % greater strength, **Figure 20.**

The project demonstrates that, particularly at the application development stage, there are major opportunities to combine commercial, production- and product-related aims with environmental protection and conservation efforts. The recognisable benefit lies:

- for the steel industry, in the substantiation of the advantages offered by steel as a material,
- for the automobile makers, in the manufacture of lower-cost, weight-optimised vehicles,
- for the consumer, in safe, more comfortable, lighter and more economical vehicles, and
- for the environment, in reduced fuel consumption and, therefore, in the conservation of resources.

102

Costs of environmental protection

The improvement in environmental protection and environmental conditions involves a corresponding increase in the investment expenditure on pollution control systems and in the operating costs incurred in the production of steel. In the past 10 years, some 500 mill. € have been invested in pollution control systems. Current operating costs at Thyssen Krupp Stahl AG in the fiscal year 2000/01 amounted to around 290 million Euros **(Figure 21)**, [ThyssenKrupp, 2001]. Water protection and conservation measures accounted for some 49%, and air pollution control for around 34% of the cost, while recycling and waste management accounted for 12%, and noise protection for just under 5 %, **Figure 22.** Whereas air pollution control used to be the most cost-intensive, water protection and conservation are today, and in future will continue to be, the largest cost factors.

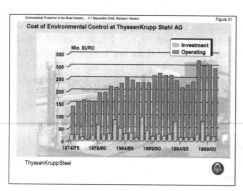
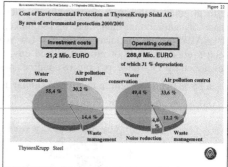

Conclusions

The findings of the Brundtland Commission and the call by the international community at the World Summit in Rio de Janeiro in 1992 have given conservation efforts new impetus. The conference in Rio de Janeiro has also strengthened the understanding that environmental protection cannot be an end in itself, but needs to be seen in an integrated manner in conjunction with other areas of policy.

A key term in this context is Sustainable Development. The steel industry can claim to have acted largely in the spirit of these modern ideas before the concept of Sustainable Development became known. This is substantiated by a great number of projects, of which the effective reduction of dust and gaseous emissions and here, in particular, the extraction of dust from steel plants, consistent recycling and waste management and, more recently, production-specific and trendsetting product developments deserve special mention.

The environmental development, position and objectives of a steel maker are described, taking the example of an iron and steel mill located in the densely populated Rhine and Ruhr region. The most important results of the enterprise's efforts can be summarised as being that:

- the specific emission of dust from the handling of solid raw materials, products and residual products has been reduced to less than 500 g/t$_{crude\ steel}$,

- a recycling rate of over 95 % has been achieved for solid, dust- and sludge-like co-products and wastes,
- the water used for process and waste-gas cleaning purposes is largely re-circulated, and
- the pollution of the air, water and soil in proximity of the works has been considerably reduced in the past few years.

General process concepts as well as appropriate guide values should serve as the basic technical standard for protecting the environment. One system possibly pointing the way for the future in this respect could be the BAT (Best Available Techniques), such as are being developed in the European Union on the basis of Directive 96/61/EG for the integrated avoidance and reduction of environmental pollution. The decisions that ultimately have to be made at a local level must, however, be reached on the basis of health protection and the ecological circumstances in each case.

The experience gained by ThyssenKrupp Stahl has shown that the environmental protection requirements are constantly being tightened and supplemented nationally and internationally, and new areas included in the environmental regulations. Examples of this are the progressive environmental policy of the European Union, as well as soil and climate protection. Steel makers are certainly well advised to gear their environmental policy to this general trend of development.

References

Agenda 21 (1992) Pressekonferenz der ThyssenKrupp Stahl AG am Agenda 21: Konferenz der Vereinten Nationen für Umwelt und Entwicklung im Juni 1992 in Rio de Janeiro. Hrsg.: Bundesministerium für Umwelt, Naturschutz und Reaktorsicherheit, Bonn, 1997.

Batz, R. (1996) Ergebnisse und Schlussfolgerungen aus dem Dioxinmessprogramm bei Anlagen zur Gewinnung von Metallen. In: Integrierter Umweltschutz in der metallerzeugenden Industrie. Preprints UTECH Berlin; Umwelttechnologieforum. Februar 1996.

Bredehöft, R. (2001) Aktuelle Entwicklung in der Abfallwirtschaft in der Stahlindustrie. Vorgetragen auf der Tagung Abfallwirtschaft 2001 am 22. Juni 2001 in Köln.

Bröker, G., Fermann, M. and Quaß, U. (1997) Identification of relevant industrial sources of dioxins and furans in Europe. Materialien No. 43. Landesumweltamt Nordrhein-Westfalen. Essen .

Our Common Ffuture (1987) The World Commission on Environment and Development. Oxford, New York: Oxford University Press (1987).

Drewes, E.-J., Engl B. and Kruse, J. (1999) Höherfeste Stähle – heute und morgen. stahl und eisen 119 (1999) Nr. 5. pp. 115/122.

EIPPC Bureau (2000)European Commission. Directorate General JRC Joint Research Centre. European IPPC Bureau. Integrated Pollution Prevention and Control (IPPC) Best Available Techniques Reference Document on the Production of Iron and Steel. March 2000, January 2001.

Hein, M., Huhn, F. and Armbruster, L. (2000) Fortschritte bei der Verringerung von Emissionen bei der nassen Kokskühlun. Stahl u. Eisen 120 (2000) No. 4, p. 103/109.

Hendricks, C., Rasim, W. Janssen, H., Schnitzer, H., Sowka, E. and Tesè, P. (2000a) Inbetriebnahme und erste Ergebnisse der Gießwalzanlage der Thyssen Krupp Stahl AG. Stahl u. Eisen 120 (2000) Nr. 2 S. 61/68.

Hendricks, C. (2002b) Kokserzeugung Heute: Bau einer neuen Kokerei in Schwelgern. Vorgetragen auf dem Umweltkongress Duisburg am 18./20. September 2000. In: Umweltschutzkongress Duisburg. Hrsg.: Amt für kommunalen Umweltschutz der Stadt Duisburg. Duisburg.

Hofherr, K., Liszio, P. and Still, G. (2001a) Improved environmental protection .through modern coke plant technology at the plant in Schwelgern. Cokemaking International 1/2001. P. 39/45.

Hofherr, K., Liszio, P. and Still, G. (2001b) Verbesserter Umweltschutz durch moderne Kokstechnik der neuen Kokerei Schwelgern. Stahl und eisen 121 (2001) No 9 p. 33/40.

Keating, M. (1993) Agenda für eine nachhaltige Entwicklung. Centre for Our Common Future. Genf.

Kersting, K., Wirling, J., Esser- Schmittmann, W. and Lenz, U. (1997) Dioxin-und Furanabscheidungaus Abgasen einer Sinteranlage an Braunkohlenkoks. stahl und eisen 117 (1997) No.11, p.49/55.

Kohler, W. (2001) Stahlerzeugung in Deutschland – eine Bestandsaufnahme. Vorgetragen auf dem Umweltkongress Duisburg am 18./20. September 2000. In: Umweltschutzkongress Duisburg. Hrsg.: Amt für kommunalen Umweltschutz der Stadt Duisburg. Duisburg.

Lindenberg, H.-U. (1999) Beiträge der metallurgischen Forschung zur Qualitätsverbesserung. Stahl und Eisen 119 (1999) Nr. 5 S. 79/85.

Masek, V. (1975) Zur Qualität fester Emissionen beim Koksnasslöschen. Staub-Reinhaltung der Luft 35 (1975) p. 87.

Nelles, L., Toll, H. and Köhler, I. (2000) CSQ-Low emission coke wet system. Cokemaking International 12 (2000) No.2, 41/45.

Nelles, L.; Toll, H.; Köhler, I.: CSQ-Low emission coke quenching system. Presented on Nov. 16, 2000 at the Internationale conference Stahl 2000, Düsseldorf . Stahl und Eisen 121 (2000).

Nelles L., Toll, H. and Köhler, I. CSQ (in press) – A low-emission coke quenching system. Stahl und Eisen.

Philipp, J.A. (1987) Retrofitting of a BOF-Steelmaking Shop`s Emission Control System. Technical Information Thyssen Stahl AG. Fachberichte Hüttenpraxis Metallweiterverarbeitung 25 (1987) Nr. 10, pp. 1100/1103.

Philipp, J.A., Görgen, R., Henkel, S., Hoffmann, G.W., Johann, H.P., Pöttken, H.-G., Seeger, M., Theobald, W., Trappe, K., van Ackeren, P., Erwe, S., Feierabend, K., Janssen, B., Maas, H., Nagels, G. and Piodrowski, H. (1987) Umweltschutz in der Stahlindustrie, Entwicklungsstand - Anforderungen – Grenzen. Stahl und Eisen 107 (1987) H. 11 S. 507/14.

Philipp, J.A., Johann, H.P., Seeger, M., Brodersen, H.A. and Theobald, W. (1992) Recycling in der Stahlindustrie. Stahl und Eisen 112 (1992) Nr. 12 S. 75/86.

Philipp, J.A. and Theobald, W. (1994a) Herausforderungen des Umweltschutzes an den Werkstoff Stahl, Vorgetragen auf der Tagung Werkstoffforschung unter Umweltaspekten, 24.-26. März 1994, Dresden.

Philipp, J.A., Eyerer, P., Erve, S., Schuckert, M., Theobald, W. and Volkhausen, W. (1994b) Stahl und Eisen 114 (1994) Nr. 11 S. 71/78.

Philipp, J.A., Theobald, W. and Volkhausen, W. (1996) stahl u. eisen 116 (1996) Nr. 11 S.99/104.

Philipp, J.A. (1997) Integrated monitoring and control – a German case study. IISI Seminar Steel on Sustainable Development, Stockholm 16.-17. Juni 1997.

Philipp, J.A. and Wemhöner, B. (1999a) Individual chamber pressure control to avoid emission of coke ovens. ECSC Workshop „Steel Research & Development on Environmental Issues"; Bilbao 10.-11. Februar 1999.

Philipp, J.A. (1999b) Reducing the emissions of dioxins from sinter plants. ECSSC-Workshop „Steel Researsch of Development on Environmental Issues"; Bilbao, 10.-11. February 1999.

Philipp, J.A. and Schulz, V. (1999c) Integrierter Umweltschutz in der Eisen- und Stahlindustrie. Vorgetragen auf der Fachtagung des Abfallentsorgungs- und Altlastensanierungsverbandes NRW, 9. Juli 1999, Hattingen.

Philipp, J.A. (2000a) Thyssen Krupp Stahl AG: Ein Unternehmen steht für den Umweltschutz.. Vorgetragen auf dem Umweltkongress Duisburg am 18./20. September 2000. In: Umweltschutz-kongress Duisburg. Hrsg.: Amt für kommunalen Umweltschutz der Stadt Duisburg. Duisburg.

Philipp, J.A. (2000b) Stand und Entwicklung des Umweltschutzes in der Stahlindustrie. Stahl und Eisen 120 (2000) Nr. 4 S. 43/51.

Philipp, J. A., Werner, P. and Wemhöner,R. (2000c) Decreasing of dioxin emissions at sinter plants. 4th European Coke and Ironmaking Congress. June 19-22,2000. Paris La Défense,France.

Philipp, J.A. (2001a) State and future prospects of environmental control .MPT International.Vol.24 (2001) No.4, p.44-55.

Philipp, J.A. (2001b) Developpment and application of the CSQ Coke quenching method. The Chinese Society for Metals. October 31-November 2, Bejing, China.

Richtlinie 96/61/EG des Rates über die integrierte Vermeidung und Verminderung der Umweltverschmutzung vom 24.September 1996. Abl. Nr. L 257/26.

Richtlinie 1999/30/EG des Rates vom 22.April 1999 über Grenzwerte für Schwefeldioxid, Stickstoffoxid und Stickstoffoxide, Partikel und Blei in der Luft. ABl. L 163 vom 29.6.1999,S. 41.

Schulz, E. (1992) Umweltschutz in der Eisen- und Stahlindustrie – Statement der Thyssen Stahl AG, IISI Jahrestagung 1991, Montreal, 07. 10. 1991. Stahl und Eisen 12, (1992) No, p. 43-51.

Schulz, E.D., Kowalski, W. Bachofen, H.J., Peters, K.-H., Wilms, E. and Land, S. (1995) Planung, Bau Inbetriebnahme und Eisen - erste Betriebsergebnisse des Hochofens 2 Schwelgern. Stahl und Eisen 115 (1995) No. 11, p. 41/54.

Stümpke, A. and Hauenstein, N. (1998) Neue Lösungen mit Stahl im Automobilbau. Stahl und Eisen 118 (1998) No.. 7, p. 75/80.

Erste allgemeine Verwaltungsvorschrift zum Bundes-Immissionsschutzgesetz (Technische Anleitung zur Reinhaltung der Luft- TA Luft) Grunddrucksache des Bundesrates 1058/01 u. Drucksache 393/02.

Theobald, W. (1995) Untersuchung der Emissionen polychlorierter Dibenzodioxine und –furane und von Schwermetallen aus Anlagen der Stahlerzeugung. Umweltforschungsplan des Bundesministers für Umwelt, Naturschutz und Reaktorsicherheit. Abschlussbericht 104 03 365/01. Düsseldorf/Berlin November 1995.

Geschäftsbericht der Thyssen Krupp AG 2000/2001.

Toll, H., Köhler, I. & Nelles, L. (2000) Das CSQ-Löschverfahren. Stahl und Eisen 120 No. 4, p.95/99.

CHAPTER 12

Development of approaches to sustainability for the metallurgical industry using mathematical modelling

Dr. Irina Butorina, Prof. Peter Kharlashin

Priazovsky State Technical University
Universitetskaya Str., 7, 87500 Mariupol
Ukraine

Summary

This chapter outlines the use of mathematical modelling in the development of sustainability approaches in the metallurgical industry.

Introduction

According to the system of environmental standards ISO 14000 series it is possible to estimate the efficiency of ecological policy applied in an industrial plant using parameters of production which take into account resource and power consumption and the formation of all kinds of wastes per ton of a product. The lower expense and waste formation levels are the more sustainable production at the given plant is and the greater its authority and competitiveness on the world market. In metallurgy the parameters of manufacture are calculated per ton of crude steel.

The values of specific parameters are influenced by the quality of raw material and technologies at all stages of metallurgical production. In common terminology, processing of raw material from iron concentrate to crude steel is called the "life cycle" of steel. The analysis of the life cycle makes it possible to define an optimum means of sustainable production.

This chapter presents the results of life cycle analysis at all stages of agglomeration, blast and steel-smelting production using an up-to-date unified method of mathematical modelling. Recommendations are made for the transition of Ukrainian metallurgical enterprises to sustainable development.

The basic equations of life cycle mathematical models were used to define specific parameters of crude steel production. They were solved together with the equations of specific resource consumption, output and structure of consumable product and solid wastes and formations and elimination of dust and hazardous gases. These equations were incorporated in a mathematical model for the calculation of the cost price of a finished product for estimating economic efficiency of production. The environmental efficiency of reconstruction was estimated using the equations of sieving of hazardous substances in the environment.

1. Mathematical model of the life cycle of steel

The equation for the calculation of specific parameters, with a special emphasis to the specific parameters of crude steel production were calculated using the following equation:

W. Leal Filho and I. Butorina (eds.),
Approaches to Handling Environmental Problems in the Mining and Metallurgical Regions, 107–117.
© 2003 *Kluwer Academic Publishers. Printed in the Netherlands.*

$$I = \frac{A_a I_a V_a}{CS} + \frac{I_{c.i.} V_{c.i.}}{CS} + I_{o-h.s.} a + I_{c.s.} (1-a) \qquad (1)$$

where A_a is a fraction of own agglomerate in the blast smelt;

V_a, $V_{c.i.}$ and $V_{o-h.s.}$ are the amounts of produced agglomerate, hot iron and open-hearth steel;

I_a, $I_{c.i.}$, $I_{o-h.s.}$ and $I_{c.s.}$ are the specific indexes calculated per ton of agglomerate, hot iron, open-hearth and converter steel;

and

$$a = \frac{V_{o-h.s.}}{CS} \qquad (2)$$

where CS is a volume of the crude steel produced.

Calculation of economical parameters of production

The equations used for the calculation of material and heat balance of metallurgical processes considered jointly at the given agglomeration charge were introduced into the mathematical model for the calculation of economic parameters, such as consumption of raw material and energy, output and structure of finished product and its cost price. The calculation is made per 100 kg of charge with the given composition. As a result of the calculation of the mass balance, the output and composition of the consumable product is defined. The input data for the calculation of a charge composition are taken from our databank containing information on the chemical composition of different metallurgical raw materials (59 items). The databank includes approximate prices of raw material used for calculating the cost price of production.

The tabulated calculation of the mass balance as a complex of chemical reactions is proposed in the given model as against the widespread practice of mass balance calculation when formation of each reactant is calculated separately. This procedure gives a picture of the metallurgical process as a whole and makes it possible to define the output of consumable products, slag, smoke gases and cumulative heat effect simultaneously. Besides, the tabulated form of calculation of the mass balance is very convenient for programming, for example, in MS Excel, the most simple programming system.

In compiling the heat balance of metallurgical reactions, some important factors of heat input, such as heat effect of metallurgical process calculated in the material balance and heat input from charge and injection, were taken into consideration. The heat consumption for the heating of charge and heat losses with gases, solid wastes and cooling water were accounted as the expense of material balance. The remaining heat losses were considered unaccounted and taken as 10 per cent of the income part of balance.

Definition of environmental parameters

A mathematical description of hazardous wastes emissions includes equations for the calculation of formation of these substances in technological processes and equations of efficiency of their elimination in various gas-processing apparatus.

Calculation of dust formation

Sinter plant. The major part of dust in the agglomeration production is produced during sintering of the charge. 95 per cent of dust emissions occur in this process.

The process of dust emission from the sintering of a metallurgical charge is characterized by two features. Firstly, the dust emission comes from two zones of the sintered layer – agglomeration zone and preheating. Secondly, each of these layers is a source of dust, on the one hand, and on the other hand they serve as filters for dust elimination. The processes of dust formation and its elimination by a layer depend on the structure of the material in each zone.

We have developed a mathematical model for this process having accepted the mechanism of a level-by-level emission of dust from the agglomerated layer. The equation (3) is used for the definition of the mass of dust emitted from dispersed layer and provides the basis of this model:

$$M_d = M_{s.f.}\left(1 - exp\left(-\frac{1.06kH^{0.5}d_m\rho_p}{\Delta P^{0.5}D^{2.5}\mu}\right) \times (0.877\times10^{-7} + 2.57\times10^{-11}t_f)\right) \quad (3)$$

where $M_{s.f.}$ is a mass of the fine fractions in a layer;

η is the efficiency of dust elimination with filtration;

D is the equivalent diameter of grains of a filter layer;

d_m is the average median size of emitted dust;

ρ_p is the density of dust particles;

H is the height of the layer;

k is the linking coefficient which allows a degree of adhesion bond of dust particles with the particles of filtering filling;

μ is the dynamic viscosity of the smoke gas;

ΔP is the negative pressure created by the vacuum chambers;

t_f is the time of filtration.

The filtration rate can be defined with sufficient accuracy using the equation of Ramzin / 1 /:

$$w_f = \left(\frac{\Delta PD}{1.2H}\right)^{1/n} \quad (4)$$

where n – is the coefficient that depends on the size and shape of the particles.

In this equation the weight of fine fractions of dust $M_{s.f.}$ in the layer of non-sintered charge can be determined with an integral curve of distribution of granulometric composition of the charge material and with the value of the critical size of particles that gain their mobility at a given filtration rate. The critical radius of particles is defined by the formula of sedimentation:

$$d_{cr} = \sqrt{\frac{18\mu w_f}{\rho_p}} \quad (5)$$

110

The specific content of fine fractions in an agglomeration layer depends on the strength of the agglomerate and the value of loads. As reasons for the agglomerate's destruction the specialists name the following / 1 /:

- presence of the fluxing components in the agglomeration charge,
- dispersed and elemental composition of the charge, and
- high rate of cooling of the agglomerate.

As a result of the mathematical processing of available experimental data on the dependence of the strength of the agglomerate on the various factors, we have obtained an empirical equation for the calculation of a mass share of fine fractions in the agglomeration layer:

$$m_{s.f.} = A_1 \times A_2 \times b^4 \times \exp(-b^2)$$ (6)

where A_1 is a coefficient allowing for a share of return in the agglomeration charge;
A_2 is a value allowing a cooling rate of the agglomerate;
b is a basicity of charge.

The value of A_1 coefficient at the share of return in the charge equal to $0 \le b \le 0.52$ can be calculated using the following equation:

$$A_1 = 0,28-0,196B$$ (7)

At the share of the charge return equal to $b>0.52$, A_1 is calculated as:

$$A_1 = 0,46B-0,68$$ (8)

In order to calculate A_2 the following equation (9) was developed on the basis of the balance equations of heat exchange of agglomerate and air filtrated through it:

$$A_2 = a_2 \times \Delta T$$ (9)

where a_2 is a constant of proportionality;
ΔT is the rate of cooling. Its value can be defined as:

$$\Delta T = \frac{T_A - T_{air}}{w_A \rho_A C_A / w_{air} C_{air} + 1}$$ (10)

where T_A is the temperature of the agglomerate;
T_{air} is the reference temperature of the air;
w_A and w_{air} are sintering rate and air filtration rate;
ρ_A and C_A are the density and heat capacity per mass unit of agglomerate;
C_{air} is the heat capacity per volume of air.

The equations show that the greater the difference of temperatures of agglomerate and cooling air the greater the sintering rate, the lower the filtration rate, the higher the strength of agglomerate and the lower the dust emission are.

The above mechanism of dust formation in agglomeration production adequately reflects the variable nature of dust ejection lengthwise as observed in practice. Here about 50 percent of the dust is ejected at the beginning of a belt, then the process of dust ejection stabilizes and at the end, when the charge goes out from the agglomeration strand, increased dust ejection is observed again.

Blast production. The main sources of dust emissions in blast production are the stock houses and the casting yard. A share in the total dust emissions (9 %) is attributed to the charging equipment and interbell spaces.

The blast furnace charge is sieved and loaded into furnace-charging carriages in the stock houses. Taking into account the factors influential on dust emission in the stock house, the specific formation of dust in this source was set from experimental data with account of decrease of fine fractions of agglomerate calculated using equation (6).

The emission of dust from the charging apparatus and interbell spaces depends on the content of fine fractions in the charge and the type of charging apparatus. Besides, the emission of dust from the interbell spaces depends on the dust emission from the blast furnace and the performance of the gas-purification channel behind the blast furnace (because the interbell space is purified with the semipure blast gas before opening of the bells) and also on the wear factors of the charging apparatus, large and small charging bells. It is impossible to take all these factors into account when compiling a mathematical model for dust emission during charging of a blast furnace. In this connection the formation of dust in these two sources was determined using the tabular data. However, in these sources the specific emission of dust will be dependent, first of all, on the main property of the blast charge, i.e. the strength of the agglomerate.

The formation of dust in the casting yard of the blast shop takes place as a result of evaporation of hot iron and slag at the overspill from the blast furnace. Experience shows that the overwhelming amount of evaporation comes from the strands of the fluid hot iron. The weight of iron evaporating in the casting yard can be defined as:

$$M_i = w_e \times F \times \tau \tag{11}$$

where w_e is the rate of evaporation, kg/sec*m^2;

F is the evaporating surface;

τ is the time of evaporation.

To calculate the rate of evaporation the equation (12) of sublimations of fluid metal was used:

$$w_e = \frac{0.5\mu^{0.5}}{(2\pi R)^{0.5}} \left(\frac{P_{vap}}{T_{vap}^{0.5}} - \frac{P_c}{T_c^{0.5}} - P_{res} \right) \tag{12}$$

P_{vap} is the iron vapor pressure at the temperature of evaporation T_{vap};

P_{res} is the residual pressure of iron vapours that is usually equal to zero;

P_c is the iron vapour pressure at condensing temperature T_c;

μ is the molecular mass of iron;

R is the absolute gas constant.

In addition to the evaporation of iron from the surface of the strand flowing through the area of the casting plant, the evaporation process also takes place at the overspill of cast iron into the steel-pouring ladle pot. In this case the surface of evaporation depends on the height of metal spray fall and its diameter. Therefore, dust formation at the overspill of metal in a metallurgical plant is a complex function of overspill rate.

The dependence on different parameters can be defined quite accurately when determining the speed of metal motion in the spray using free-falling velocity and an equivalent sectional area of the spray. In this case the equation for calculation of the mass of the evaporated metal will be the following:

$$M_m = 1.68 \times W_p^{0.5} \pi^{0.5} H^{0.75} g^{0.25} \tau^{0.5} w_e \qquad (13)$$

According to this equation, the overspill rate applicable to the minimum value of evaporation depends on the weight of poured metal, distance from the point of overspill to the bottom of the ladle pot, elemental composition of metal and its temperature. So, it is necessary to allow for particular conditions choosing the optimal speed of metal drainage that reduce dust emission to the environment.

Steel-smelting production. The main sources of dust in the steel smelting plants are the places of fluid metal overflows and the process of steel smelting per se. Dust formation at the overspill of hot iron and steel in the steel smelting plants is the same as above legitimacies. Calculating dust formation at the spray overspill of cast iron and steel using the equation (13) has given good concurrence with the experimental data.

The formation of dust during processing of hot iron into steel in steel smelting furnaces depends on the technology and the aggregate used for the smelt of steel. The level of increase in specific dust formation in steel smelting productions is as follows: converter, open-hearth and electric steel smelting production. Both the process of evaporation and the process of granular materials' submission are the reasons for dust formation in a fluid bath of metal. The surface of evaporation significantly influences the process of evaporation. In its turn it depends on a free mirror of metal, impact velocity of an oxygen spray with fluid metal, amount of generated bubbles etc. It is rather difficult to take all factors influential on dust emission from steel smelting aggregates into account. So, the specific emission of dust from them was based on experimental data.

Calculation of hazardous gases' formation

In mathematical models of basic metallurgical production the mathematical descriptions of the formation of the main toxic gases of metallurgical production such as CO, NO_x, SO_2 were given.

Formation of CO and SO₂

Sinter plant. In agglomeration production the toxic CO and **SO₂** gases are formed during incineration of solid fuel. In mathematical models of the formation of these gases we have proceeded from topochemical mechanism of solid fuel incineration and interaction of sulphur oxides with lime stone allowing the following assumptions:

— the fuel particles in the agglomeration layer are of equal size and spherical shape;

— the process of fuel incineration proceeds in a limited spherical zone of incineration the radius of which is permanently reduced during the process;

— the width of the incineration zone in the agglomeration layer varies with the height. It increases in the upper part of layer and decreases in the lower part at the same rate while it is constant in the central part where it is defined as:

$$h = 2{,}3410^{-4} C_\% O_\% \, exp\left(\frac{0{,}5 * 10^{-6} \Delta P}{r} \right) \qquad (14)$$

where $C_\%$ and $O_\%$ are percentages of carbon in the charge and oxygen in the filtrated air;

r is the radius of fuel particles;

ΔP is the negative pressure under agglomeration strand.

Using the above assumptions the equation (15) for the definition of the specific formation of CO per kg in production of one ton of agglomerate that shows the dependence of this value on the basic technological parameters of the process is:

$$CO= \frac{10}{a\rho_c}(4,65\ C_\%\rho_c-1,75\frac{w_f\rho_{o2}O_{2\%}(H + h)}{w_sH}) \qquad (15)$$

where F and H are the area and height of sintering;

w_f and w_s are the rates of air filtration and charge sintering;

ρ_{O2} is the density of air;

ρ_c is the charging dense of charge;

a is a share of output of the finished product.

In mathematical models of the specific output of SO_2 the balance of sulphur oxides formation as a result of fuel incineration and its absorption with lime as the element of charge, was compiled. As a result the equation (16) for the calculation of the amount of SO_2 that is emitted with smoke gases during the time period $d\tau$ was developed:

$$dSO_2 = \frac{F_A\rho_c}{100}(\frac{2h\ C_\%k_rC_{O2}S_\%}{100r_T\rho_T}-\frac{3H_cCaO_\%r^2k(C_{SO2}-C^*_{SO2})}{r^3\rho_{CaO}})d\tau \qquad (16)$$

where ρ_{CaO} is the density of lime;

$S_\%$ and $CaO_\%$ are percentages of sulphur in the fuel and lime in the agglomeration charge;

C_{SO2} and C^*_{SO2} are concentrations of SO_2 in the reaction zone and the balance concentration of SO_2 for this reaction;

k_r and k are constants of the rate of topochemical reaction of fuel incineration and the rate of interaction of sulphur oxide with lime;

r_T and r are radiuses of fuel incineration area and area of interaction of sulphur oxide with lime. This equation is solved jointly using equations of variable height of the incineration zone (h) and the zone of non-sintered charge H_c.

Blast and steel-smelting plants. The main sources of CO emissions in blast plants are the torch where the excess of blast gas is incinerated, and gas heaters, and in steel smelting plants – the furnaces themselves. It is difficult to calculate the output of CO from this source. It was taken from reference material. The process of CO formation in the gas heaters cannot be calculated because it is influenced by various factors. The CO output from steel smelting production is calculated using material balance equations with downstream use of the coefficient of after-burning of this gas in the torch. The content of sulphur oxides in the products of blast gas incineration and in gas heaters was defined according to the structure of blast gas calculated with the material balance of blast smelt. The output of SO_2 in the steel smelting plant was defined in the same way.

Formation of nitric oxides

We have proposed a simplified way of calculating nitric oxides output in metallurgical production based on the computational ratio of thermodynamic balance of nitrogen and oxygen. As a result a

114

semi-emprical equation for calculation of nitric oxide output during incineration of any fuel was developed

$$NO_X = 1{,}2\alpha L_T M \exp(-2{,}1H_2)\left(\sqrt{kK_b} - 0{,}25K_b(1+k)\right) \qquad (17)$$

where NO_X is the weight of nitric oxides (in kilograms)   formed at incineration of a given weight of fuel (M); H_2 is the percentage of carbon in the reaction zone; α is the coefficient of air flow rate; L_T is the theoretical oxygen flow rate (in moles) for oxidizing a mass or volume unit of given fuel; k is the coefficient equal to ration of nitrogen percentage in oxidizing mixture to oxygen percentage; K_b is the constant of the thermodynamic balance of oxygen and nitrogen interaction reaction. Applying equation (17) we see that nitric oxide emission in agglomeration, blast and steel smelting processes and in fuel incineration correlates to experimental data quite well.

Calculation of gas purification devices and sieving of emissions

In order to include the catching of emissions with gas purification devices, equations making it possible define the efficiency of gas purification performance, the value of hydraulic resistance and water consumption were introduced into the mathematical model. The equations for the calculation of the inertia chamber, cyclone, injector and Venturi scrubber, electric precipitator and bag filter were taken from well-known methods / 3 /.

To define the influence of emissions from various metallurgical sources on the quality of atmospheric air the method for calculation of sieving OND-86 / 4 / was used. The maximum concentration of hazardous substances in the ground layer and the distance from the source of emissions to the point of maximum concentration were defined.

Results of research and recommendations for sustainable metallurgical production

The main results of computational research with the developed mathematical model applying the technical and economical parameters of Mariupol metallurgical works as the input data, are presented in tables 1-3 and figures 1-4. The economical and environmental parameters of steel production in EU operations are given in comparison.

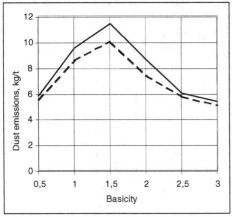

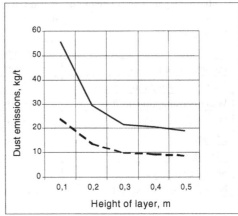

Fig. 1. Dependence of specific dust emissions on basicity of charge

—— share of return is 25 %;

– – – share of return is 35 %

Fig. 2. Dependence of specific dust emissions on the height of layer

—— share of fine fractions is 0.02;

– – – share of fine fractions is 0.006

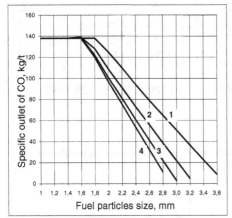

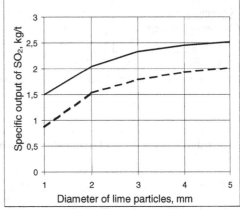

Fig. 3. Dependence of CO output on the fuel particles' size

Negative pressure in vacuum chambers

1 – 400 mm hg; 2 – 600 mm hg;

3 – 800 mm hg; 4 – 1000 mm hg

Fig. 4. Dependence of SO_2 output on the diameter of lime particles

—— Diameter of lime particles is 1 mm

– – – Diameter of lime particles is 2 mm

Table 1. Dependence of economical and environmental parameters of steel production on the type of iron ore concentrate

Parameter	Type of iron ore concentrate			
	Kerchensky	NKGOK	YuGOK	Lebedinsky
Percentage of iron, %	48,6	61,5	64,5	66,3
Energy consumption, GJ/t	27,3	22,27	22,07	21,68
Water consumption, m^3/t	30,26	29,93	29,92	29,9
Dust emissions, kg/t	8,38	7,15	7,13	7,05
CO emissions, kg/t	26,33	26,16	26,15	26,14
NO_x emissions, kg/t	1,77	1,5	1,46	1,42
SO_2 emissions, kg/t	3,26	0,798	0,775	0,759
Non-utilized wastes, kg/t	866,3	550	546	522

Table 2. Dependence of parameters of production on the share of open-hearth production*

Parameter	EU enterprises	Ukrainian steel producers					
		Share of open-hearth steel					
		0	0,2	0,4	0,8	0,9	1
Energy consumption, GJ/t	19,5-20	22,2	21,4	20,6	19,7	18,9	18,1
Water consumption, m^3/t	0,4-20	29,9	28,3	26,7	25,1	23,5	21,9
Dust emissions, kg/t	0,24-0,5	7,15	7,11	7,06	7,02	6,98	6,94
CO emissions, kg/t	2,09-9,44	26,1	25,6	25,2	24,7	23,8	23,8
NO_x emissions, kg/t	0,7-1	1,51	2,28	3,06	3,83	4,61	5,39
SO_2 emissions, kg/t	1,2-2,5	0,79	0,87	0,94	1,02	1,09	1,16
Non-utilized wastes, kg/t	9-118	550	503	457	411	365	319

* It was assumed that the steel smelting production consisted of open-hearth and converter plants

Table 3. Dependence of parameters of open-hearth production on the share of scrap

Parameter	Share of scrap in open-hearth smelt				
	0,4	0,3	0,2	0,1	0
Energy consumption, GJ/t	17,6	19,8	21,5	24,5	26,4
Water consumption, m^3/t	21,6	23,7	25,8	28,0	30,2
Dust emissions, kg/t	6,8	7,7	8,5	9,3	10,1
CO emissions, kg/t	47,6	54,8	62,0	69,3	76,5
NO_x emissions, kg/t	5,0	5,1	5,2	5,3	5,4
SO_2 emissions, kg/t	1,5	1,6	1,7	1,9	2,1
Non-utilized wastes, kg/t	340	374	427	481	534

The following conclusions can be drawn from the computational data analysis:

1. The parameters of steel production in the Ukrainian metallurgical industry do not meet international standards at all stages of metallurgical processing. They provide evidence of the unsustainable development of steel production in the Ukraine.

2. A significant improvement in the parameters of steel production can be achieved by reducing resource consuming and wasting agglomeration and blast production by using high quality raw material and an optimal share of scrap in the steel smelting process (not less than 18 %).

3. The modernization of agglomeration production aimed at increasing the quality of the product, reducing energy consumption and pollution of the environment should be developed in the following directions:

- reconstruction of sintering machines so as to reduce air inflow (inflow coefficient ≤ 1.4) and to organise a system of smoke gases recycling;
- change of sintering technology with transition to sintering of two layers of charge with different basicity (0.5 and 2.5) at the increase of return share in charge up to 35 percent;
- installation of electric precipitators downstream from all sources of emissions.

4. To increase environmental parameters of blast production it is necessary:

- to use high quality agglomerate with low content of fine fractions;
- to install bag filters in the aspiration system of under-bunker chambers;
- to reconstruct existing systems of gas purification downstream of the interbell spaces;
- to install covers with a system of aspiration and high effective gas purification at the drain spouts;
- to provide high performance incineration of blast gas in the gas heaters;
- to provide utilization of blast gas with liquidation of incineration torches.

5. To improve economical and environmental parameters of steel smelting production it is necessary:

- to close open-hearth plants. If there is no electric steel smelting plant at the plant it is necessary to reconstruct sintering plants before closure of open-hearth plants. Otherwise it will be necessary to increase energy consumption and waste smelting of hot iron if scrap processing capacities are reduced, hence worsening environmental and economical parameters;
- to implement mobile mixers;
- to remove cracks in the walls and windows of converter plants and install the suction-and-exhaust ventilation with purification of gases in electric precipitators;
- to collect and utilize converter gases.

6. If specific water consumption is to meet international standards it will be necessary to transfer to the closed cycle of water-supply and not use wet gas purification at the sintering and steel smelting plants.

Finally, funding for reconstruction of enterprises can be obtained from an economical use of energy and resources and from balancing trade policy.

References

Vegman, E.F. (1974) Theory and technology of sintering. Moscow: Metallurgy, 28 p.

Philipov, S.I. (1967) Fundamentals of metallurgical processes. Moscow: Metallurgy, 230 p.

Stark, S.B. (1990) Gas purification apparatus and metallurgical devices / Handbook, 2nd edition, Moscow: Metallurgy, 405 p.

Normative documentation OND-86. Moscow: Hydromet, 1987, 93 p.

CHAPTER 13

Prospects of Structural Reconstruction of the Mining and Metallurgical Complex - on the Basis of Use of Iron Ores Produced by Hydraulic Mining through Boreholes

Prof. Nikolay Lukianchikov, Dr. Luisa Gagut

Scientific Centre in the Russian Research Institute
of Economics of Mineral Raw Materials and Subsoil Use (VIEMS)
3-ya Magistralnaya st. 38
Moscow, 123007
Russia

Summary

This short chapter offers an overview of some of the possibilities available via using hydraulic mining through boreholes.

Introduction

In the Kursk Magnet Anomaly (KMA) area, over 60 billion rich iron ores have been found. As for their quality – the iron content is more than 60-66%, there is low silica content and a low level of other admixtures – these ores are natural superconcentrates and they can be used, after insignificant beneficiation, as raw material for direct metal smelting and in the powder metallurgy sector. Until recently there has been virtually no demand for rich KMA ores because of the great depth at which they occur (400-600 meters) and because of their significant water content. The Russian Institute of Mineral Raw Materials (VIMS) and other scientific and industrial organisations in Russia have developed and tested industrially the non-traditional method of hydraulic mining of iron ores through boreholes, which helps make commercial exploitation efficient. For rich KMA ores, the cost of production using the hydraulic mining through boreholes (HMB) method is 2-3 times lower than that of open pit mining and is 5-7 times lower than that of underground iron ore mining. Capital investment is a few times lower than in traditional mining operations and can be repaid in 2-3 years. The smelting of steel from the metallized product helps reduce the amount of harmful admixtures emitted to the air by 70%.

This could be done by the consortium of Russian and Ukrainian mining and metallurgical enterprises with the participation of specialised banks and large energy concerns (Gazprom, for example). In this case it would be possible to make use of various economic advantages provided by existing legislation in Russia and the Ukraine, such as, for example, production-sharing agreements, agreements of financial and industrial groups etc. It is possible to locate a suitable plant for the production of metallized briquettes from KMA ores using the HMB method in a special economic zone of priority investment development in the Cis-Azov region. This will help to solve the problems of the thorough structural reconstruction of the metallurgical complex in Mariupol and secure ecological improvements in the Sea of Azov basin, which is important both for Russia and the Ukraine.

The mining of low-grade ores and their subsequent beneficiation accounts for the overwhelming proportion of iron ore production in the CIS countries. The relatively low grade of iron ores is the

119

W. Leal Filho and I. Butorina (eds.),
Approaches to Handling Environmental Problems in the Mining and Metallurgical Regions, 119–121.
© 2003 *Kluwer Academic Publishers. Printed in the Netherlands.*

reason for the high resource- and power-intensive indices involved in obtaining the end metal products; these indices are 1.2-1.4 times higher than those in industrially developed countries. This results in high environmental contamination in large metallurgical centres, such as the complex in the Cis-Azov region.

At the same time, over 60 billion rich iron ores have been found in the Kursk Magnet Anomaly (KMA) region. After beneficiation these ores are significantly superior to iron ores – the iron content is more than 60-66%, the silica content is low as is the level of other admixtures. In essence, these ores are natural superconcentrates and they can be used, after insignificant beneficiation, as raw material for direct metal smelting and in the powder metallurgy sector.

Up until lately the rich KMA ores have been virtually non-requested because of the large depth of their occurrence (400-600 meters) and significant water content. Russian Institute of Mineral Raw Materials (VIMS) and other scientific-and-industrial organizations in Russia have developed and tested industrially the non-traditional method of hydraulic mining of iron ores through boreholes, which helps put them into efficient commercial exploitation. For rich KMA ores, the cost of their production using the hydraulic mining through boreholes (HMB) method is 2-3 times lower than that of the open pit mining and is 5-7 times lower than that of the underground iron ore mining. The capital investments are a few times lower than those at enterprises of traditional mining and they can be repaid in 2-3 years.

Research conducted by VIMS, Central Scientific Research Institute of Ferrous Metallurgy and Gipromez shows that the maximum ecological and economic effect can be attained when the HMB method is used in the non-coking metallurgy blast-furnace process stage, owing to the exclusion of agglomeration and coking as it is this which is the greatest cause of environmental contamination, rather than using the HMB method in traditional metallurgical processes. The smelting of steel from the metallized product helps reduce the amount of harmful admixtures emitted to the air by 70% [1].

Metal production using the "direct reduction – electric smelting" scheme is cheaper than the "coking – blast furnace processing – steel smelting" process. Capital costs of production of 1 t of steel under the classical scheme are 450 – 600 dollars while the costs involved in the direct reduction of metal are 300 – 400 dollars [2]. However, the development of the latter method is hampered by limited availability of high-quality iron ore material which can be used for the process of direct reduction of iron. In the USA, the average profit of small metallurgical mini-plants, where the metallized products from Latin American countries rich in ores are used, is 148.6 dollars per ton of steel products shipped, while the profit of large plants employing the traditional technological process of metal smelting is 60.33 dollars per ton [3]. Over the last 40 years the proportion of such mini-plants in the USA has grown from 10-15% in 1960 to 45-50% today.

The annual productivity of mines where the HMB method is used can vary significantly: from a few hundred thousand tons of ore to a few or even dozens of millions tons. Quite significant (billions of tons) reserves of rich friable ores from each of the KMA iron ore deposits found at considerable depth make it possible to design HMB plants of any possible productivity, which is in fact limited by real investment opportunities.

The commercial development of prospective KMA deposits using the HMB method can be realized in stages. The initial annual productivity is 200-400 kt with an annual increase of production to 4-5 Mt achieved mainly with the help of internal financial sources. When constructing an HMB plant at such a level of productivity it is necessary to drill 4 technological 219-mm boreholes, one or two of which are intended for production, the other two or three being for pumping purposes. The pumping and compressor stations, pipelines for the compressed air and water, a pipeline for the pulp, settling basins and other structures securing the implementation of HMB should all be located at the mine site.

In essence, it is possible to locate a plant designed for metallized briquette production at any metallurgical plant where electric steel smelting is used and natural gas is available on site. The construction of the plants designed for metallizing of raw materials and the appropriate temporary closing-down of coking-blast furnace capacities will have significant positive ecological effects.

From the transportation point of view, it is advisable to use KMA ores produced by the HMB method on Ukrainian territory. This might be done by the consortium of Russian and Ukrainian mining and metallurgical enterprises with the participation of specialised banks and large energy concerns (Gazprom, for example). In this case there is a possibility to use various economic advantages granted by the existing legislation in Russia and the Ukraine, e.g., the production-sharing agreements, agreements on the financial-and-industrial groups, etc. It's possible to locate a plant for production of metallized briquettes out of KMA ores produced with the help of the HMB method within the area of a special economic zone of the priority investment development in the Cis-Azov region. Thus, it will help to solve the problems of the deep structural reconstruction of the metallurgical complex of Mariupol and ecological improving of the Azov Sea basin, which is important both for Russia and the Ukraine [4].

Notes

1. Астафьев Н.П., Гагут Л.Д. Приоритетное направление в добыче железных руд КМА. – Горный журнал, 1995, № 1, с. 13.
2. Astler E. (1991) Potential for small scale Ironmaking // Steel Times International, 1991, v. 15, № 1, pp. 18, 19.
3. US Minis embark on a new phase // Steel Times International, 1992, v. 16, № 3, p. 23.
4. Гагут Л.Д. СНГ: Новый путь развития в XXI веке. – М.: "Русь", 2000 – 384 стр.

CHAPTER 14

Ecological problems of the geological environment caused by activities of the metallurgical plant "KCM" (Bulgaria)

Dr. Aleksey Benderev, Dimitrov, Prof. Elka Pentcheva, Elitsa Hrischeva

Geological Institute - Bulgarian Academy of Sciences,
Acad. G. Bonchev str. 24
Sofia 1113
Bulgaria

Summary

Since 2001 a team of scientists from the Geological Institute, BAS - Sofia, Bulgaria, BRGM, Orleans, France and the University of Antwerp, Belgium has been working on a project entitled "Dynamics, evolution and limitation of heavy metal water pollution in the Plovdiv region (Bulgaria), financed by the NATO Programme SfP. The main research objects are soils, sediments, waste materials and waters in the region of "KCM" smelter Plovdiv. The main activities during the first year period of the Project consisted in examination of the existing information, collection and analysis of soil, sediment and water samples, and initial monitoring observations. The first results show that the pollution of the area is located/localized mainly around the "KCM". As a result of emissions released from the plant, soils from a large area are polluted with Pb, Zn and Cd. The waste waters from the "KCM" discharging directly into Chepelarska River account for pollution with Pb, Zn and Cd of the surface waters of the river. Leaching of the soluble phases of the exposed solid waste materials has caused local pollution with Cu, Mn and Fe of the ground waters. Ground waters at close proximity to the KCM production site are contaminated mainly with Pb, Zn, Cd, Cu and As. Phases controlling the migration of heavy metal pollutants have been detected. The acquired results will be used as a basis for further GIS data integration and prognostic modelling of pollution evolution and measures for limitation.

Introduction

In areas of mining activity and in the metallurgical industry the diffusion of technologically polluted surface and meteoric waters (from aerosol clouds) is today endangering groundwater resources —a situation that is likely to worsen in the future. The problem of pollution by heavy metals is of particular importance, with the integral effect and chemical speciation of the metal ingredients being a significant hazard to human health. One of the plants that has a harmful effect on the environment in Bulgaria is "KCM" smelter situated to the south of the town of Plovdiv (Fig.1). The smelter was established in 1961 for processing of lead-zinc ores. Since that time the activity of "KCM" has caused considerable environmental problems in the region, which is one of the most fertile in Bulgaria.

For ecological risk assessment of the region a project entitled "Dynamics, evolution and limitation of heavy metal water pollution in the Plovdiv region (Bulgaria)" was proposed by scientists from the Geological Institute, BAS – Sofia in collaboration with BRGM, Orleans, France and the University of Antwerp, Belgium and accepted for financing by the NATO Programme "Science for Peace". The principal objectives of the project are as follows:

W. Leal Filho and I. Butorina (eds.),
Approaches to Handling Environmental Problems in the Mining and Metallurgical Regions, 123–137
© 2003 *Kluwer Academic Publishers. Printed in the Netherlands.*

Fig. 1. View of the smelter "KCM"

- To establish the distribution speciation and interactions of heavy metals in soil surface and groundwater in the Plovdiv region;
- To determine the effect of localized and diffuse pollutants on water and soil quality;
- To establish changes in heavy metals and As concentration over time;
- To develop a 3D GIS modelling of surface and ground water pollution sensitivity and GIS based pollution-risk assessment tools and methods;
- To develop a predictive coupled geochemical and transport computer programme for studying the evolution of pollution with the aim of limiting its harmful effects.

The objective of the present chapter is to present the results obtained during the first year period as well as the activities planned for the accomplishment of the project.

Geological-hydrogeological characteristics of the region

Surrounded by the mountains of the Rhodope Massif and the Sredna Gora, essentially composed of ancient igneous and metamorphic silicate rocks, the Plovdiv plain is filled with sedimentary rocks and alluvial deposits of the Maritza River and its tributaries (including Chepelarska River) (Kouzhoukharov at al., Explanation note..., p. 41). The actual study area (more than 200 km^2) is located south of the Maritza River and consists principally of the left-bank terrace of the Chepelarska River up to the town of Asenovgrad (Fig. 2). The soils and alluvia are layered, slightly alkaline, carbonaceous, and poor in humus (1-2%).

Quaternary and Pliocene sediments form a general aquifer where ground waters are isolated in some places, but in most cases the waters have hydraulic connection due to the similar lithological composition of the sediments (Antonov and Danchev, Ground waters ..., pp. 208-229). Sediments consist of boulders, gravels, sands, clayey sands and clays. The thickness of the sediments (from 10 to >300 m) increases from South to North. The depth of the ground water level with respect to the surface varies from 2.2–2.5

m near the Maritza River to 25-26 m in the southern part of the area. The flow direction is to the north and to the Maritza River, respectively. The natural resource of the water flow passing through the study area is more than 3000 l/s. This water is the main source of the drinking and industrial water supply. The surface water has been used only for irrigation. Both deep and shallow aquifers have been used. A few pumping stations were built, the biggest of which (a line of wells) provides about 400 l/s and supplies "KCM". Each of the other pumping stations has 1 or 2 wells which have been used for different purposes.

Main sources of pollution associated with "KCM"

The main sources of pollution in the area are the mining products from neighbouring regions, as well as some imported ore concentrates processed in "KCM". The smelter recovers about 55,000 t of zinc and 40,000 t of lead annually. The "KCM" also produces around 8000 t/yr of lead/antimony alloys, and recovers cadmium and precious metals as by-products. "KCM" is considered the main pollutant of the area with heavy metals – lead, zinc, copper, cadmium and arsenic.

There are several ways in which the environment is polluted:

- Pollution caused by the industrial activity within and at a close proximity to the production site;

- Dispersion and deposition of the emissions released from the smelter. The aerosols, contaminated by heavy metals and arsenic up to 30 times above the Maximum Permissible Limit, are dispersed over some 3-4 km around the "KCM". Emissions have caused permanent pollution of the soils and intermittent pollution of the agricultural yield from a large area surrounding the plant. Over the past few years the influence of pollution from emissions has decreased due to the installation of purification filters. At present contamination of the soils is considered residual;

- Transportation of hazardous elements by industrial effluents. For transportation of the waste waters from "KCM" a canal was constructed. A large part of the canal is in a concrete bed, with the exception of its end section where waste waters discharge into the Chepelarska River. There is no system for purification of the waste waters;

- Effect of the smelter waste disposal. There are two dumping sites for storage of the solid waste materials. One of them was used till 1970 and the other one is currently functioning.

Planned activities

The following work packages were defined within the project for the achievement of the formulated objectives:

WORK PACKAGE 1. Hydrogeological characteristics of the studied region including elucidation of the conditions of ground water formation and their distribution, filtration parameters, hydrodynamic situation and elaboration of conceptual groundwater model of the studied region.

WORK PACKAGE 2. Field and laboratory hydrochemical investigation consisting of: sampling of waters and suspended matter; "in situ" water analysis for unstable compounds or parameters (Eh, pH, conductivity, hydrocarbons, sulphides); laboratory analyses of 57 trace elements, including Pb, Cd, Zn, Cu, As; the complete background hydrochemical characterisation will be based on 20 major ingredients; computation of the important chemical speciation of the metal pollutants, together with the forms and migration of the other trace and major elements; study of the hydrogeochemical metal behaviour, based on associations, correlations and ratios, coefficients of anomaly and toxicity, distribution of the pollutant elements and their environmental impact in the Plovdiv region.

WORK PACKAGE 3. Geochemical and mineralogical study of soils, sediments and waste materials including: soil grid sampling on low and high density networks; complex geochemical-mineralogical characterization; application of Sequential Chemical Extraction for selected samples; grain-size analysis.

WORK PACKAGE 4. Geographic Information System (GIS) integration and modelling. A satellite imagery and Digital Elevation Model will be used as the basis. GIS integration of all collected and/or interpreted data relevant to the area of study. Application of 3D GIS tools to understand and model the pollutant dissemination pathways.

WORK PACKAGE 5. Experimental modelling of interphase processes including: experimental modelling (simulation of interphase processes under various conditions and factor influences) for determining probable natural geochemical barriers to prevent and halt the infiltration of harmful mobile forms of metallic pollutants in the underground hydrosphere. Study of sorption, desorption, ion-exchange, precipitation and redox processes (with their dynamics and kinetics) as well as leaching properties.

WORK PACKAGE 6. Computer elaboration and modelling. Design of a Specific Chemical Simulator (SCS) taking into account the main geochemical characteristics of the region. Coupling the SCS with the general hydro-transport modelling tool MARTHE (BRGM's property).

WORK PACKAGE 7. Interpretation and synthesis. Predictive interpretation of the evolution of the metal migration under various forms and pathways for possible inactivation. Synthesis of all results (and tools) for the design of a complex "anti metal pollution" strategy, with know-how and forecasting.

The relation between the individual work packages and the activities of the participants in the project are presented in Fig. 3.

Field work and methods

Geological and hydrogeological field trips were carried out for field inspection and after that sampling according to the accepted in the Project scheme have started (Fig. 2).

Water sampling

Initial general sampling of water sources in the area - about 100 samples - was carried out. The sampling included: surface waters along the Chepelarska River, before and after influx of the waste water from the "KCM" canal and along the Maritza River; factory waste waters, as well as surface drainage waters from stock and waste piles and other solid pollutants; groundwater sources using an irregular sampling network according to the position and influence of atmospheric, localized and diffuse pollution. Subsequent seasonal samplings in the most endangered areas have also started. Denser surface and underground sampling grids (about 4x30 samples) were used in order to

127

determine the temporal chemical water dynamics and to select monitoring points for the 4D model. So far three seasonal samplings have been carried out

One more seasonal samplings is planned and after that regular monitoring of selected surface and underground points will start. About 12x10 samples will be taken over a 1-year period on a monthly basis in order to refine the temporal chemical alterations and pollutant movements.

Soil, sediment and waste material sampling

A low-density grid sampling of soils at 2.5 km intervals followed by higher density subsampling was carried out. Seventy point samples were taken at a depth of 0-30 cm. The sampling was performed according to the planned scheme (Fig.2). Stream sampling included collection of two types of sediments: surficial (0-2cm) sediments at the sediment-water interface, taken from waste water canals of "KCM" and Agria along the Chepelarska River streambed and the Maritza River; deeper sediments taken on two profiles. A total of 20 sediment samples was collected. Different types of waste materials were taken from old and new dump sites. Temperature and pH of fumarols were measured "in situ".

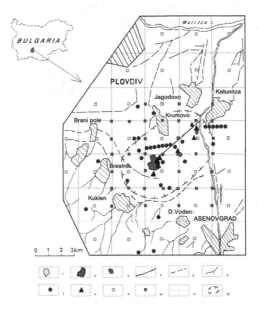

Fig.2. Scheme of the region with main water sources and the soil sampling points:
1. Populated area 2. KCM 3.Dumps 4 Drainage canal of KCM 5. Canals 6. Rivers 7. Main water sources 8. Monitoring boreholes drilled in 1999 9. Soil sampling points (low density) 10. Soil sampling points (high density) 11. Soil sampling net 12. Provisional area for detailed study

Methods

Water samples were taken according to ISO5667 Water quality International Standards requirements. Water analyses "in situ" for non-stable components (CO_2, HCO_3^-, CO_3^{2-}, dissolved O_2, pH, Eh and conductivity) were carried out. Laboratory chemical and physico-chemical analyses of all water samples for the identification of 24 physico-chemical parameters, including Ca^{2+}, Mg , Na^+, K^+, Cl^-, SO_4^{2-}, NO_3^-, NO_2^-, F^-, NH_4^+ were carried out according to Standard Methods for the Examination of Water and Wastewater (APHA/AWWA/WEF 1992) and BDS Standards for Drinking and Industrial Waters (BDS 2825-83). All water samples were analysed for the following 34 elements: Al, As, Ba, Cd, Co, Cr, Cu, Fe, Mn, Ni, Pb, Sb, Zn, Hg, Be, B, V, Sn, Mo, Se, Ag, Te, Tl, Ti, U, Ca, Li, Mg, Sr, Bi, Ga, In, K, and Na. The analysis was carried out on HR-ICP-OES Jobin Yvon ULTIMA 2000, (at UIA-MiTAC and GEOLAB), in conformity with ISO 11885.

Soil samples were air-dried and < 2 mm fraction sieved for X-ray diffraction and chemical analyses. Soil pH was measured in a soil/distilled water suspension in the ratio 1:2.5. Grain-size analysis comprised wet sieving and settling procedures. X-ray diffraction analysis for the determination of the mineralogical composition of bulk samples and < 2 μm clay fractions were carried out using Siemens D 500 diffractometer, Cu Kα radiation, 40 kV, 30 mA. Concentrations of the major elements were determined by XRF analysis. Concentrations of trace elements- Pb, Zn, Cu, Ni, Co, Cr, Cd were analysed by AAS in bulk samples after complete dissolution with NHO_3, $HClO$, HF and in exchangeable phases after extraction with $MgCl_2$.

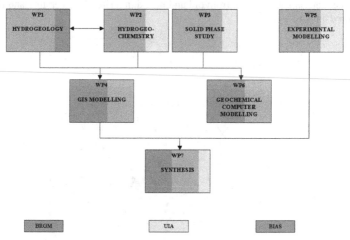

Fig.3. Relation between individual work packages.

Preliminary results

An essential part of the activities before and after the project began involved collection of information concerning the geology, hydrogeology and ecology of the region. The main part of the information used was provided by the Ministry of Environment and Water and consisted of unpublished reports. Some data concerning pollution of soils was given by Hinov and Hajtondjiev (Heavy metals.....) and Hristova et al. (Geochemical characteristics ..., pp. 12-14).

The collection of additional hydrogeological information (as concerning the bedrock in the southern part of the ditch, other boreholes in and around the region, lithological data and etc.) as well as other data continues. Climatic data from 3 national meteorological stations in the region are available. Hydrological measurements are carried out at two river gauges stations – on the Chepelarska River (south of the region) and on the Maritza River in the town of Plovdiv. It was established that the main sources of pollution in the region apart from "KCM", are the settlements, the agricultural works and industrial enterprises. The strongest pollutant among the communities is Assenovgrad with its 52,000 citizens. There is a sewage system in the town, but a water purification station has not been constructed. The sewage is poured out directly in Chepelarska River. More than 50 industrial plants are to be found in this region. The most dangerous of these as far as ground water is concerned is AGRIA, a factory producing fertilizers and a factory of the food and wine industry in Asenovgrad.

A special database called WATMETAPOL DATABASE was developed (Petkov, in press; Petkov et al., in press) for the needs of the Project, based on the information collected. The database will be used as a basis for future investigations and hydrogeochemical modelling.

Smelter waste disposal

During the current investigation some new facts concerning waste materials were established (Atanassova and Kerestedjian, in press). Diverse mineral species were found to form abundant aggregates around specific gas exhalations (fumaroles) in the metallurgical clinker heaps of the "KCM" smelter. These fumaroles are generated by a process of burning the coke inside the heap mass. The high temperatures at combustion points cause decomposition of the glassy clinker mass and uplift of chemical components forming new mineral phases on the heap surface. The measured gas temperature varies from 40 °C in the oldest parts of the heap to 211 °C in hot spots on the new dump. The pH of the gas also varies from 3 in relatively exhausted parts of the heap to below 1 in the hot spots. The intensity of the burning process depends on the composition of the clinker at the specific location, the amount of water, drainage conditions etc.

For the clinker produced at present the amount of coke and the concentrations of all the elements both in the clinker and its water solutions are below the maximum permissible limits (MPL). However, clinkers produced only 4-5 years ago are significantly above the allowed limits. The old dump materials, after almost 30 years of coke burning and natural rain leaching, show concentrations about 10 times higher than MPL. The worst situation is with the materials 5-20 years old (on the new dump). They show concentrations of Cu 800 mg/l, Zn 270 mg/l. Rains play an important role in the overall environmental picture here. On the one hand, each time a sufficiently long period of rain comes, it dissolves and takes away all the newly formed mineral products; on the other hand, the water drained inside the heap intensifies the chemical processes there and new quantities of recent phases are formed on the surface over and over again.

Soils

The predominant type of soil in the area is alluvial with highly variable mechanical composition. Depending on their grain-size composition, soils range from loamy sand, loam, to silty loam and in a few cases clay loam. Soils in the area are slightly alkaline with pH values in the range 6.2-7.7. Analyses of the mineral composition of soils show that clay minerals and quartz represent the dominant constitution. The amount of clay minerals is up to 65% in finer-grained soils, while quartz is found in greater amounts (up to 50%) in sandy soils. Other registered minerals are K-feldspar (0-24%), plagioclases (0-12%), calcite (0-28) and dolomite (0-8%), showing uneven distribution within different soil types. The analysis of < 2 μm fraction shows that illite is the dominant clay mineral in soils (41-73 %), kaolinite is second in abundance (16-42 %) and chlorite (0-19 %) and smectite (0-16 %) are present in minor amounts.

130

The distribution of heavy metals and As in soils is influenced by natural and anthropogenic factors. The analysed elements could be divided into two groups based on the relative importance of each one of these factors on their distribution The first group comprises Cd, Pb, Zn, Cu and As. Concentrations of Cd, Pb and Zn in soils throughout the region show considerable variability (Cd - <0.1-77 mg/kg; Pb – 25-4400 mg/kg; Zn – 43-5200 mg/kg) with higher values registered in the soils around "KCM". Lead in concentrations exceeding Bulgarian maximum permissible limits (MPL) was registered in soils at a distance of about 6 km, and Zn and Cd – about 4 km SE of "KCM". The distribution of Pb concentrations is presented in Fig. 4. The distribution of Zn and Cd concentrations shows similar features. Concentrations of Cu (14-352 mg/kg) and As (0-79 mg/kg) are elevated only at close proximity to "KCM", while in other cases they do not exceed MPL. The dispersion and deposition of elements are affected by the prevailing wind directions. The distribution patterns of concentrations of this first group of elements clearly indicate pollution from the emissions coming from the "KCM" smelter. The maximum of Pb, Zn and Cu near the new dump shows local influence of the solid waste materials on soil pollution. The second group of elements comprises Cr, Ni, Co, Mn and Fe. Their concentrations are less variable throughout the region (Cr – 37-335 mg/kg; Ni – 16-170 mg/kg; Co – 6-27 mg/kg; MnO – 0.05-0.26 wt%, Fe_2O_3 – 2.3-8.65 wt%). The elevated values are not predominantly located in soils surrounding "KCM" but are also detected in soils from a wider area covering the southern and southwestern part of the region (Fig. 5). It may be suggested that in addition to input from "KCM", the elements have been naturally supplied within materials coming from the erosion of nearby metamorphic rocks. An indication of this is the elevated concentration of Mn, Fe, Cr and Ni in soils developed on metamorphic rocks in the area.

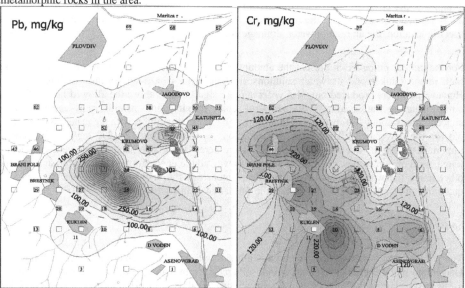

Fig. 4. Distribution of Pb concentrations in soils

Fig. 5. Distribution of Cr concentrations in soils

The analysis of elements bound as exchange cations in soil phases show that the potentially most mobile and harmful element is Cd (there is up to 50% of Cd in highly polluted soils in exchangeable phases). Other potentially mobile elements are Cu, Zn and Mn.

The distribution of heavy metals deep in the ground was investigated by analyses of core samples from wells situated within and at close proximity to "KCM" as well as from wells near "KCM" canal at the point where the waste waters discharge into the Chepelarska River. In the first group of wells it was registered that concentrations of elements are higher close to the surface and decrease downwards. In most cases, concentrations of heavy metals remain elevated down to a depth of 1.5m indicating that infiltration of pollutants by rain waters is taking place in this highly contaminated area. In wells located close to the canal an increase in concentrations of Zn, Pb and Cu was registered in sediments at about 2m and 4.5 m depth intervals.

Waters

The waters in the region are of $Ca(Mg)HCO_3$ (SO_4) type, mainly HCO_3-Ca and HCO_3-SO_4-Ca. Abundance of Cl^-, NO_3^- and SO_4^{2-} varies considerably between wells. The presence of considerably high values for ammonium and nitrate ions as well as increased concentrations of nitrites and sulphates imply groundwater contamination by infiltrating surface water which has been polluted by leakage from waste heaps and sewage discharges.

Analyses of trace elements in waste waters coming from "KCM" show highly elevated concentrations of Pb (0.36-1.54 mg/l), Cd (0.16-3.83 mg/l) and Zn (4.8-157 mg/l) (Velitchkova et al., in press). The analyses of Chepelarska River water indicate that the greatest risk with regard to heavy metal pollution, is represented by the water after discharge of the waste waters from the canal into the river. The seasonal variations of concentrations of Zn, Pb, Cd and Cu (Fig. 6) show that the water quality is highly influenced by the quality of the waste waters discharging at any particular moment. In most cases the effect of the waste waters diminishes downstream.

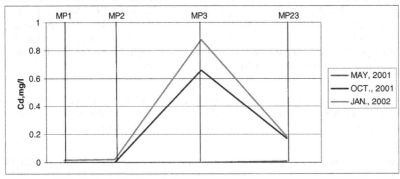

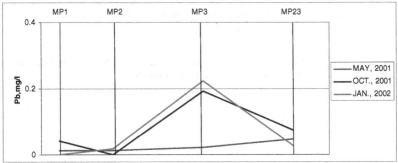

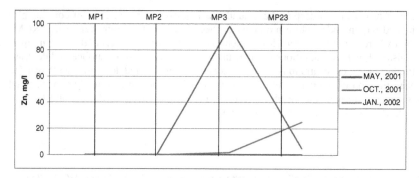

Fig. 6. Concentrations of Pb, Zn and Cd in the waters of the Chepelarska River at four points
(MP1 – south of Asenovgrad, MP2 – north of Asenovgrad, MP3 – after discharge of waste water canal, MP23 – at the
confluence of the Chepelarska River with the Maritza River).

The ICP values of the water samples are summarized and compared in accordance with the existing standards in Bulgaria – Ecological limit, being a warning characteristic and Pollution limit the point above which the waters are considered polluted (Table 1). A pollution of Fe,Hg, Mn, Cd, Ni and Al of the unconfined underground waters in the range of 24 to 36% has been determined. This is also valid to a lesser extent for Se and Tl. This contamination is of regional distribution and is presumably of geogenetic origin.

A significant proportion of the host rocks consist of boulders and rubble of magmatic and metamorphic rocks from Rhodopes province. The highest concentrations of Fe, Mn and Cd were registered in a local area around "KCM" and are a result of "KCM's" activity. The considerably lower standard limits of some toxic elements such as Hg, Cd, Se and Tl in comparison with other elements should be pointed out. About 6 to 8 % of the water samples show Cu, Zn, Pb and As contamination, which is localized around "KCM" and is a result of its activity.

The data from the seasonal monitoring show that the ground waters are contaminated in a local area around "KCM" with heavy metals – Pb, Zn, Cu, Cd, As , emitted from the smelter. Common tendencies regarding distribution of concentrations were determined for all the elements:

- The areas with maximum pollution are localized within or at close proximity to the "KCM" industrial site;
- The distribution of maximum concentrations of each element in ground water coincides with the respective maximums in soils, but is more local in character;

The extent of the pollution is variable during the seasons with maximum values registered in the spring months when rates of recharge usually reach the highest levels.

Table 1. Data on chemical elements in underground water composition [mg/l]

Element	Number of samples	Typical statistical values				Standards for Bulgaria		Number of samples above the standard limit	
		Min	Max	Average	St. Dev.	Ecological limit	Pollution limit	Ecological	Pollution
Ag	50	0	0.0006	0.0001	0.0002	0.003	0.01	0	0
Al	50	0	1.7752	0.1494	0.3811	0.2	0.5	20	12
As	103	0	0.9542	0.0272	0.1534	0.01	0.03	18	8
Ba	50	0.01	0.7000	0.1104	0.1067	0.2	0.5	3	1
Be	50	0	0.0006	0.0001	0.0001	0.0002	0.002	4	0
Cd	103	0	7.1570	0.0877	0.7072	0.001	0.005	35	27
Co	50	0	0.0183	0.0009	0.0026	0.005	0.1	0	0
Cr	50	0	0.0334	0.0044	0.0085	0.005	0.05	11	0
Cu	103	0	1.4624	0.0419	0.1867	0.03	0.1	35	6
Fe	50	0	10.5949	0.6763	1.6568	0.05	0.2	24	17
Hg	50	0	0.0690	0.0040	0.0126	0.0005	0.002	43	16
Mn	103	0	11.2552	0.3200	1.4008	0.02	0.05	35	27
Mo	50	0	0.0218	0.0010	0.0034	0.005	0.05	2	0
Ni	50	0	0.2500	0.0444	0.0690	0.02	0.1	24	13
Pb	103	0	1.2562	0.0348	0.1407	0.03	0.1	13	7
Sb	50	0	0.0125	0.0006	0.0018	0.005	0.02	13	0
Se	50	0	0.0194	0.0020	0.0042	0.0005	0.005	24	6
Sn	72	0	0.0325	0.0017	0.0048	0.01	0.1	17	0
Te	50	0	0.0300	0.0032	0.0041	0.01	0.04	1	0
Ti	50	0	0.0850	0.0044	0.0127	0.2	0.8	0	0
Tl	50	0	0.1490	0.0069	0.0210	0.001	0.01	32	5
U	50	0	0.0600	0.0154	0.0114	0.01	0.1	33	0
V	50	0	0.0165	0.0021	0.0027	0.05	0.2	0	0
Zn	103	0	16.2081	0.4819	1.9438	0.2	1	20	6

Fig. 7 illustrates Pb-distribution in ground water in evolution as compared with its distribution in the respective soils.

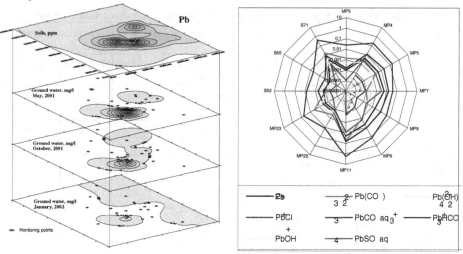

Fig. 7. Seasonal distribution of Pb concentrations in ground waters compared with its distribution in soils

Chemical speciation

An important thermodynamical study for establishing the soluble species of existence and migration of heavy metal pollutants (against the background of the elucidation of the major component forms) was carried out by Pentcheva et al. (Chemical speciation...., in press). The mobile migration forms of the pollutants define their biological activity and toxicity. The WATEQ 4F programme was applied. About 60 chemical species (CS) for the HMP were found in all the waters, about 40 others – for the major components. More than 200 figures were established for the mineral phases, controlling the migration in each of these waters and those that are closest to the equilibrium (very different for ground and surface waters) were selected. The established CS show a pronounced diversity, with different relative weight and accumulation possibilities illustrated for Pb (Fig.8). In the ground water a generalized rate for the Pb species importance can be described as follows: $PbCO_3 > PbHCO_3^- > Pb^{2+} > PbSO_4 > Pb(CO_3)_2^{2-} > PbOH^+ > Pb(SO_4)_2^{2-}$. In surface waters as well as in waste waters different carbonate species predominate (together with hydroxide and sulphate forms). Cd in ground waters predominates in carbonate forms, in waste and river waters – in cationic Cd^{2+} form. Cu varieties are difficult to systematize, but most significant are carbonate and hydroxide species (of various weight) and cationic Cu^{2+} in the most polluted waters. Important for Zn is Zn^{2+} cationic form, especially in the most contaminated waters, where a similar range is likely to be observed. Example: $Zn^{2+} > ZnHCO_3^+ > ZnSO_4 > ZnCO_3 > Zn(SO_4)_2^{2-} > Zn(CO_3)_2^{2-} > ZnOH^+ > ZnCl^+$. In the case of Mn the simple cations Mn^{2+} and also $MnHCO_3^+$ predominate. Fe is found in the form $Fe(OH)_4 > Fe(OH)_2$ in surface waters, while in ground waters the rate is often as follows: $FeOH^+ > Fe(OH)_2 > Fe(OH)_3 > Fe(OH)_4 > Fe^{3+}$. The ratio: $HAsO_4^{2-} > H_2AsO_4^-$ is important for the principal As forms.

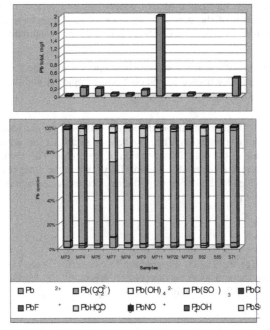

Fig. 8. Pb speciation compared to the total concentrations

Analyses of suspended matter collected on membrane filter 0.45 μm have been carried out, and different mineral and chemical forms were identified by EPMA investigation. In waste waters twenty trace elements were established (including Hg, Ta, Se, Y) and the following composition was determined: 85% ZnO, 6% FeO, 2% PbO, 1% CuO.

Conclusions

The "KCM" plant is the major pollutant with Pb, Zn, Cd, Cu and As in the area under study. There are several ways in which pollution of the environment occurs:

- Industrial activity causing pollution within and at close proximity to the production site. The results show that infiltration of rain waters through highly contaminated soils have caused pollution of ground waters with Pb, Zn, Cd, Cu and As here.

- Dispersion and deposition of emissions released from the "KCM". Metals and metalloids in the emissions are largely in the form of particulate matter. Soil analyses show that, due to the emissions, a large area around "KCM" is polluted with Pb, Zn and Cd. The "KCM" accounts only for local accumulation of Cu and As in soils. The regional character of distribution of elevated concentrations of Mn, Fe, Cr, Ni and Co throughout the studied area suggests their predominantly natural source.

- Release of pollutants with the waste waters. The waste waters from the "KCM" discharge directly into the Chepelarska River and account for considerable pollution of the surface waters of the river with Pb, Zn and Cd.

- Exposed waste materials. Infiltration of rain waters leaching soluble phases of the wastes has caused local contamination of the ground waters with Cu, Mn and Fe. Dissemination of solid particles from the tips has caused contamination of soils only at close proximity to the new dump.

The presence in solution of heavy metal mobile species and their secondary modification, bioactivity and toxicity are determined by the chemical complexity and the good migration ability of their simple cationic, carbonate, hydroxide, sulphate etc. forms

The intermediate results of the work on the project are sufficient for elucidation of the effect of the "KCM" plant on environment pollution. They can be used as a basis for further GIS data integration and prognostic modelling of the pollution evolution of the region. The end users of the project are acquainted with the current ecological situation. They expect the final results will facilitate the planning of future activities for limitation of pollution. Economic and administrative measures will be recommended to this end.

Acknowledgements

The authors acknowledge the NATO Programme "Science for Peace" for the financing of the project. We also thank the end users of the project, especially the Ministry of Environment and Waters and Government of Plovdiv region, for their support.

References

Antonov, H. and Dantchev, D. (1980) Ground waters in Bulgaria. Sofia, Technika, p. 260.

Atanassova, R. and Kerestedjian, T. (2002) Heavy metal remobilisation in an old clinker waste heap in the vicinity of KCM smelter - Plovdiv, Bulgaria, Proc. Depo Tech – Abfallwirtschaftstagung – Leoben, Austria, Nov., 20 - 22, 2002.

Hinov, G. and Fajtondjiev, L. (1977) Heavy metals contents in the soils in the region of KCM "D. Blagoev" – Plovdiv, Soil science and agrochemistry, XII, 5, 1977.

Hristova, J., Kujkin S., Kosturkova P. and Dantcheva, N. (1994) Geochemical characteristics of the technogenous soil pollution in areas of the mining and metallurgical industry, Minno Delo i Geologia Journal, 2, 1994, pp. 12-14.

Kouzhoukharov, D., Kouzhoukharova, E. and Marinova, R. (1992) Explanation note to the geological map of Bulgaria 1:100000. Sheet Plovdiv. KGMR and PGPGK, Sofia, 1992, p. 41.

Pentcheva, E., Velitchkova, N. and Van't dack, L. (in press) Chemical speciation in solution and in suspension of heavy metal pollutants in waters (Plovdiv region, Bulgaria), 8-th FECS Conference on Chemistry and Environment. Athens, Greece 31.08–4.09, 2002, Abstract in Environmental Science and Pollution Research.

Petkov, I. (2002) Application of the Information systems (Gis) for hydrogeological assessment of heavy metal pollution of ground waters around KCM – Plovdiv, Minno Delo i Geologia Journal, 2002.

Petkov I., Spassov, V., Benderev, A. and Hristov, V. (in press) Hydrogeological background and heavy metal contamination of groundwater in industrial zone: Plovdiv-Assenovgrad (Bulgaria), Proc. XVII Congress of Carpathian - Balkan Geological Association – Environmental Geology, 1 - 4 Sept. 2002, Bratislava, Slovakia.

Velitchkova, N., Pentcheva, E. and Daskalova, N. (in press) Investigation on the concentration levels of heavy and toxic elements in environmental materials from Plovdiv region, Abstracts of International Workshop: "Solubility phenomena – Application for Environmental Improvement", Varna, Bulgaria, July 22-26, 2002, (in press).

CHAPTER 15

OMENTIN - information network about mining and environmental technologies

Prof. Janos Foldessy[1], Balazs Bodo[2]

[1]University of Miskolc, 3515 Miskolc Egyetemvaros, Hungary
[2]Geonardo KFT, 1031 Budapest Keve u 17, Hungary

Summary

The European mining industry faces increasing challenges to meet environmental requirements and to convince local communities of the need and benefits of its existence. Communities and residents near mine sites are increasingly concerned about the use of the different mining and processing technologies. They need to know the scientific background of these technologies, their impact on the environment and the risks involved. This information, which normally comes from the authorities or companies concerned, should be detailed, simple, transparent and unbiased. To access the public and provide them with such information improved techniques and a multinational network are needed. In this network mining professionals and environmentalist should find a common language and platform to discuss the benefits and hazards of mining and formulate joint opinions. The OMENTIN projects, described in this projects, aims to establish and develop this platform.

Introduction

One of the most traditional industries in Europe is ore mining. This is especially true of our region, Central-Eastern Europe, where mining tradition goes back to Roman times. Ore mining exploits mineral resources. Mineral resources may create wealth for communities. The exploitation of mineral resources may also create long lasting environmental problems and risks. The utilization of these resources thus has to be evaluated from both an economic and an environmental point of view. Whether or not to exploit the mineral resources is linked to the question whether the risk is acceptable or not, i.e. if this activity presents hazards to the human or natural environment.

No human society in developed countries can live without the use of metals, which surround us in the form of widely diversified products, from cars to CD players. The growing use of metals runs parallel with the growing concern regarding the environmental impacts of ore mining, and tendencies to restrict or completely ban this industry. Our view is that ore mining should be sustainable in the long-term future, as sustainable as the use of metals in the different products which are considered as necessary for our life. Rendering European ore mining unsustainable would restrict our metal consumption to imported raw materials and simply export the environmental impacts of ore mining to the developing countries of other continents.

The participants and objectives of the OMENTIN project

Mining accidents cause damage to the environment, but also initiate efforts in technological developments to increase safety. The Baia Mare accident in Romania and the Aitik accident in Sweden in 2000 have shown that accidents do occur irrespective of the development level of the

139

W. Leal Filho and I. Butorina (eds.),
Approaches to Handling Environmental Problems in the Mining and Metallurgical Regions, 139–143.
© 2003 *Kluwer Academic Publishers. Printed in the Netherlands.*

country's mining industry. These accidents were the starting point of a new initiative to make technical information more understandable and available to a wide public. This initiative has come from different European countries, and has finally been transformed into a project which has won the support of the European Commission.

The OMENTIN (Ore Mining and Environmental Technologies Information Network) is a 3-year-project in which mining professionals, geologists, environmentalists work together in assessing and evaluating and explaining hazards linked to ore mining.

The project was started after the Baia Mare and Borsa and Aitik environmental mine accidents. The initial participants in the project are:

- Geonardo KFT, Engineering and consulting company, Hungary
- CENTEK, Research and development company, Sweden
- University of Leoben, Institute for Ecosystem Analysis, Austria
- University of Baia Mare, Department of Geology and Mining Engineering, Romania
- Regional Environmental Centre, Environment Foundation, International

The staff on the teams are geologists, mining professionals, media experts and environmental scientists. They monitor, collect and interpret mining related data. Omentin aims to establish co-operation between environmental and mining experts, enhancing the objectivity of the information.

The first step of the programme is to assess the state of the art in the European mining industry and mining waste management. The traditional methods, open pit mining, underground mining, leach mining are the main alternatives for the extraction of ore. In several cases a combination of two methods was used during the lifetime of the mine. The largest European underground operations produce about 10 million tons per year, open pit operations reach the 20 million ton mark. Most of the ore mines in Europe are small by world standards.

While several of the European ore mines use traditional ore processing facilities, the newly opened mines are equipped with the best available technologies. There are basic elements of ore processing like comminution, gravity and magnetic separation, flotation, biological oxidation, leaching other solvent extraction methods, roasting. The different processing routes are based on these components. Base metals mines in Europe use the traditional grinding-flotation technologies; the leaching, solvent extraction methods are not used extensively. Many of the former uranium mines have used some version of leaching extraction.

There are traditional mining centres (like Rio Tinto in Spain, Cornwall in England, Erzgebirge in Germany, Garpenberg in Sweden, Banska Stiavnica in Slovakia, Rosia Montana in Romania, Majdanpek in Yugoslavia etc) which have been in operation for several centuries or several thousands of years. The central parts of Europe play a minor role in the mining industry today. The weight at present is on the northern (Sweden, Finland), eastern (Bulgaria, Romania) and western (Spain, Portugal) peripheries. Although the regulations regarding new mines are becoming increasingly more stringent, several important projects are in the development stage in Sweden, Finland, Romania and Bulgaria.

In the first step, the entire present-day European mining industry has been reviewed, and data collected on the ore mines which were (or have been) active in the last 15 years, i.e. from 1987 to

2002. An easy-to-use database is constructed from these data, which contains and exhibits these data by countries and by commodities.

A review of environmental regulations, as a second objective, helps to understand the legal and social aspects of these technologies. Technical and legal preparatory work is underway in several cases in the EU to provide regulations and recommendations for mining waste management and safe mine closures. Our project creates an up-to-date source of information on these recommendations.

The third part of the work programme deals with public awareness of mining science and technology. The European public is concerned about the vicinity of mines, the hazard potential of ore processing, and unforeseen consequences of mine closures. It is necessary to raise public awareness to understand industrial techniques, accept their existence in our surroundings, and enforce the mitigation of their environmental effects. It also means being prepared to deal with a problem if it occurs. Raising public awareness may mean enforcing safe and clean mine techniques, while understanding and containing their potential danger to the environment. It also means permitting mining activities and requiring assurances to safe and environmentally acceptable modes of closure at the end of their lifespan.

The final part of the work programme covers the evaluation of the Baia Mare and Aitik accidents. The evaluation aims to revise the consequent measures rather than the causes of the accident, which have been studied in detail previously by other groups. (Bergström and Bodo 2001).

Tools, working methods used by OMENTIN

During the OMENTIN programme several forms of media are tested for use in dissemination of mining and environmental technology information. Scientific evaluation and assessment of mining data is our basic method of data acquisition. Information is collected and systematized from public sources, publications, Internet releases and scientific studies. Main parameters are selected and compared to characterize the size and type of mining related hazards. Data is standardized for classification and analysis purposes.

The Internet website of the project www.omentin.org is our most important tool and the place where we publish information. The site is active and has been periodically refreshed since the starting date of the programme, May 2001. The main pages of the website are now also translated into national languages.

A newsletter is published periodically (2-4 times a year). In the newsletter the ongoing events, information about the progress of our work, new contacts, etc are given. The newsletter now appears in a growing number of national languages and is distributed electronically among environmental groups and mining professionals in a growing number of countries.

A database of active ore mines, OREMINE has been compiled and attached to the website. This database includes basic data on all European ore mines. Geographical location, main mineral commodities, production figures, mining method, processing technology are the main entries of the database, which includes the data of ore mines which were active, closed or opened in the period covering the last 15 years.

Online workshops are being organised on the Internet. This tool serves to discuss recent or forthcoming events of interest to the mining industry, such as the publication of the Cyanide Code. We plan to extend the use of this tool since the Internet is the most common media used and reaches a wide range of people in a cheap and effective way.

Technical publications are prepared which review the ore mining industry, waste management methods and environmental regulations related to mining. The first of these reports was completed in May 2001, and gave a cross section of the industry titled Ore Mining, Processing and Waste Management in Europe. The report is targeted at people with technical education, but no specific geological or mining training. It also serves as information to journalists. The reports are downloadable from the Internet.

Using these sources as background, popular science publications are written and released. These foster understanding of several mining terms as well as of risks and hazards related to mining. 'Ore mining – with or without' is the title of our first release. The publication is also downloadable from the website.

Workshops are planned to train environmental NGOs on mining related environmental risks. The first of such workshops is planned to be held in 2003.

Our final objective, and possibly the most powerful tool is the growth of the network. Since its start new partners joined us from Norway, Turkey, the Czech Republic and Romania. We are inviting other new partners who intend to participate in our activity. There is, however, still much to be done to convince environmental organisations of the usefulness of the contacts which can be built through this network. The gap created by the lack of information and lack of confidence can be closed gradually by supplying clear and understandable information. On the other hand, environmentalists cannot be unaware of society's needs in terms of mining production. We believe that by transforming scientific data into information accessible to a wide public, our work will assist both sides in hammering out a working compromise.

Sustainability of ore mining in our region

Ore mining still has a significant share in the national economies of Yugoslavia, Bulgaria, Macedonia, Albania and Greece. The Carpatho-Balkan region has outstanding ore resource potential, with large-scale exploration projects in Romania (Rosia Montana Au, Rosia Poieni Cu), Greece (Sappes, Viper, Au) and Bulgaria (Krumovgrad, Au).

Sustainability means exploiting resources without damaging future generations' living environments and opportunities of accessing natural resources. Both ends of this sentence are important. Exploiting natural resources is no longer a purely economic question. Instead, a complex of environmental, social and technical questions have to be answered positively when considering the economics of a successful ore exploration project.

The Central-Eastern European countries face an important decision regarding their mining industry. Maintaining the mining industry is possible only if methods and management techniques are revised and updated. When it is a matter of preserving our water reserves and natural habitat from irreparable damage caused by large scale mining accidents, obviously the decision should be to restrict mining. The introduction of stringent quality assurance methods during production and replacement of outdated mining and processing equipment are the technical prerequisites of the future environmental sustainability of ore mining in our region.

An open approach must be adopted and all relevant environmental information must be made public if we wish to achieve social and political acceptance of ore mining. This is perhaps a harder and more time- consuming task than solving technological problems.

Mining engineers and ore geologists can play an important role in solving the technological, environmental and social problems. Our experience, which is an alloy of earth sciences and engineering technologies, could serve as a tool to create a common language for technologies and environmentalists in conflict over mining issues.

Abandoned waste dumps, marginally economic mines, and new installations in old infrastructural environments have had to face growing requirements regarding environmental and natural protection. Not complying with these requirements has resulted in a growing number of environmental accidents (e.g. Baia Mare, Borsa in Romania, 2000) and latent pollutions (e.g. Velez smelter, Macedonia, 2001). Such environmental accidents triggered continent-wide protests and campaigns which affect not only the plants in question, but also the entire industry.

Acknowledgements

The OMENTIN project is fully supported by the European Commission's programme for "Raising Public Awareness of Science and Technology".

References

Bergström J., Bodo B. (2001): Raising Public Awareness through Mining-Sector Networking. in: Securing the Future International Conference on Mining and the Environment. July 2001 Skelleftea Sweden, Proceedings. Vol.1. pp. 38-44.

CHAPTER 16
Tools and methods for using clean technologies in the mining and metallurgical regions of NIS countries

Prof .William Zadorsky

Ukrainian Ecological Academy of Sciences
Ukrainian State University of Chemical Engineering
str.Komsomolskaya, 41/43 – 23
Dnepropetrovsk 49000
Ukraine

Summary

This chapter presents and discussses some clean technologies that may be used in mining and metallurgical sectors.

Introduction

Cleaner Production (CP) strategy and tactics, based on the systematic approach, take into account the following factors:

- All ecological problems should be solved in co-operation with unified comprehensive planning
- Ecologising the economy assumes modernization of objects which are real or potential pollutants of the environment
- The success of ecologising processes demands professionals skilled in the theory and practice of ecologising, cleaner production and ecological management
- The creation of a civilized ecological market is a necessary prerequisite for ecologising the economy and sustainable development of the country.

Only mutually balanced simultaneous comprehensive tackling of the three tasks (economical growth with simultaneous improvement of ecological conditions and dealing with social problems) will allow a progressive CP strategy to be realized. The system analysis shows strong interaction and feedback among these three factors of CP strategy.

CP algorithm using the systematic approach

The system approach described is the basis of the proposed strategy for the systematic reduction of environmental pollution. It assumes that, before problems on methods of industrial waste conversion or utilization choice can be solved, it is necessary to consider questions of systematic reduction of environmental loads at production level. It is very important to realize economically justified variants of removal or essential waste reduction by selectivity of main process raising at the lowest hierarchical object levels.

A CP algorithm will look at the following sequence of actions:

W. Leal Filho and I. Butorina (eds.),
Approaches to Handling Environmental Problems in the Mining and Metallurgical Regions, 145–162.
© 2003 *Kluwer Academic Publishers. Printed in the Netherlands.*

DECOMPOSITION. Hierarchy level determination and technical system decomposition. The analysis of the initial information including inspection of industrial manufacture, with the purpose of determining the various levels of hierarchy (for example, manufacture – plant item - installation –apparatus or machine - contact device - molecular level)

IDENTIFICATION of an initial level. Revealing the bottom level of hierarchy limiting from the point of view of pollution of the environment. Definition of limiting/limitings hierarchical level/levels. Herewith reasonable move on the hierarchical stairway from top to bottom and use methods of expert evaluations

SELECTIVITY & INTENSITY INCREASE. Increase of selectivity and intensity of actual technological stages of processing at a limiting level of hierarchy. Choice of CP methods depending on the limiting level scale (defining size) corresponding with parameters of influence method to the system from the database

The system approach that is demonstrated in Table 1 connects the levels of a system with the frequency order, dimension order, concepts and methodologies and with the tools and methods needed for cleaner production. There should be a correlation between one particular level in the hierarchy and the methodology of characterization, assessment or influence used at this level.

Two aspects of the system approach to cleaner production are addressed: firstly, the vertical hierarchy. This implies that any subsystem of a system may be regarded either as a lower-level system in relation to the level above or as an upper-level system in relation to the level below. Secondly, a correlation between a level in a hierarchy and the methodology of characterization, assessment or influence used at this level: this aspect does not seem to have been sufficiently covered previously and deserves more detailed attention. The tools used to analyse, study and influence an object should match the respective dimensions and frequency in the order of magnitude.

Table1. The system approach, concepts and methodologies connected with tools and methods for cleaner production.

№	Tier of system	Frequency order	Dimension, Order, m	Concepts and methodologies	Tools and methods for Cleaner Production
1	Man-nature-technology	$0.1\ yr^{-1}$	10^7	SD	Systematic approach
2	Consumption sector	$1\ month^{-1}$ to $1\ yr^{-1}$	10^4	SD, LCA, MM, FCA, ST, WM	Recycling of the goods as raw materials
3	Manufacturing	$0.001\text{-}0.01\ s^{-1}$	10^2	SD, EM, MM, FCA, ST, LCA, P2, ES, WM	Industrial symbiosis as a basis for management of secondary materials and energy. Flexible synthesis systems and adaptive equipment to embody them
4	Plant	$0.01\text{-}0.1\ s^{-1}$	10	SD, WFT, MM, ST, P2, ES	Process engineering for high throughput to cut processing time and reduce by-products and wastes. Local neutralization of pollution
5	Plant item	$0.1\text{-}1\ s^{-1}$	1	MM, P2	Block-modules equipment. Multifunctionality
6	Apparatus or machine	$1\text{-}10^4\ s^{-1}$	1	MM	Recycling flow of the least hazardous agent taken in excess over its stoichiometric value. Flexible synthesis systems and adaptive equipment to embody them.
7	Work assembly	$1\text{-}10^4\ s^{-1}$	$10^{-3}...10^{-6}$	MM	Contacting phases controlled heterogenization. Chemical-separative reactions: removal of reaction products at the moment of their formation. Synthesis and separation in an aerosol to increase intraparticle pressure and reaction rate. Self-excited oscillation of reacting phase flows at frequencies and amplitudes matching. Multiplicity of resources and energy use
8	Molecular level	$10^5...10^8\ s^{-1}$	$10^{-9}...10^{-12}$	Physics, chemistry	Excess of nontoxic reagent over its stoichiometric value. Minimising of the process time. Selectivity increasing with a change of the physical-chemical parameters.

NOTES EM - environmental management, WM - waste management, FCA - function-cost analysis, LCA - life cycle assessment, MM - mathematical modelling, SD - sustainable development, ST - solutions theory, WFT - waste-free technologies, P2 - pollution prevention, ES - energy saving.

Tiers 1 –4: "Man-nature-technology". Tools and Methods for Sustainable Development (SD)

A significant problem in the Ukraine is the fact that economic, social and ecological factors are treated in isolation from one another when it comes to accepting decisions made with regard to policy, planning and management. This has significant influence on the realization of the concept of sustainable development in the country and, above all, on its industrial and agricultural production. The Ukraine needs environmentally sustainable economic and social development.

As widely acknowledged, the concept of sustainable development includes three aspects: ecological, economic and social. Underestimating any one of these three components results in a distortion of this equilateral triangle and to a deviation from the strategy of sustainable development. If, in assessing economic forces, social and ecological aspects are underestimated, this is bound to have a negative effect on the stability of development, for it is impossible to ensure that conditions are improved for the following generation if economic improvement is not accompanied by both a reduction in the technogenic load per capita and solutions to the social problems of a community. Just as there cannot be by end in itself a reduction of technogenic loads per capita, and, means, the decision of ecological problems nor can be end in itself, as in a limit it would result in a return to a primitive community where the ecological balance was assured. Thus, only a counterbalanced simultaneous complex decision involving all three tasks of sustainable development - growth of the economy with a simultaneous improvement of ecological conditions and tackling of social problems - will make it possible to realize this progressive strategy. The system analysis shows strong interaction, direct and feedback, between the three factors of sustainable development mentioned. In this connection the strongest parameters determining the stability of development, are those which influence at least two of the three factors of sustainable development. Cleaner manufacturing has an influence on the economic and ecological characteristics of a system and, consequently, can be regarded as one of the basic factors ensuring the sustainable development of the system.

As it is frequently impossible to reduce the level of negative influence industry has on the environment without changing technological processes, nature conservation should be directed either at improvement of the existing situation or at introducing predominantly new technological processes, directed not only on, to decide helpful problems, but also on the protection of the environment. Thus, it is necessary to develop a system which provides a quantitative evaluation of the ecologisation of industrial objects and of the extent to which ecosystems have been destroyed, in order to ensure acceptance of justified decisions directed at the ecologisation of industrial and agricultural manufacture and at rehabilitation of areas polluted by toxic substances. An important problem is the integration of processes of acceptance of decisions in the field of the environment and development and the improvement of transfer systems and analytical methods. Comprehensive measures which evaluate the consequences of the decisions in economic, social and ecological terms are necessary, not only on the level of the individual projects, but also on the level of policy and programmes. The analysis should include an evaluation of costs, profits and risks.

The next step is the formation of a concept of sustainable development as a synthesis of the following problems:

- economic development;
- population adoption to and positive participation in technological and social changes;
- ecologisation of the economy (transformation of agriculture and industrial complexes to reduce technological loads on man and on the environment) on the basis of system-structural analysis.

A lot of scientists are suggesting that an economic-environmental-social model be developed and employed for the purposes of the country's sustainable development. This is a very complex and time-consuming approach that may not be appropriate at this time of industrial restructuring, privatization and other complex processes occurring in a collapsed national economy. An alternative approach is put forward, which is applicable at both national and regional level. Instead of mathematical modelling and optimization, it is necessary to use a systems approach and decision theory techniques.

What makes our approach different from the mainstream international cleaner production (CP) movement is the desire to abolish the dominating "black-box" techniques. Instead of regarding a production facility as no more than a given set of benign inputs and polluting outputs, we insist that one should seek the best ways to bring about a prospective cleaner object within the "black box".

As the major principle of cleaner economy, a systems approach is adopted that deals with perfecting any nature-technology system at various hierarchic levels, from environmental pollution sources to consumers, and takes into account the interactions and mutual effects of all important components. This type of analysis will reveal relationships between the ways to improve processes and the challenges of risk management and nature conservation. The main task is, therefore, to harmonize the nature-technology relation and, ideally, to engineer high-performance systems featuring desired environmental characteristics at each hierarchic level, so that the favourable environmental background is not impaired and, where possible, even restored.

The following are the basic assumptions underlying the cleaner economy concept for transfer economy countries:

- At this time of deep economic crisis, the economic and environmental challenges must be met simultaneously, in keeping with the strategy of a cleaner economy.
- The development towards a cleaner economy must focus not on consuming, but rather on perfecting those entities that are actual or potential polluters.

The success of a cleaner economy policy will be largely determined by the availability of professionals well trained in the theory and practice of „economy clean-up" and environmental management.

No cleaner economy will be possible without creating an environmentally sound market. These strategical principles require tactical measures if they are to be are pursued. Such measures are applicable to any industry and include:

- no waste due to improved selectivity;
- neutralizing wastes directly at the source, rather than at the exit;
- flexible technologies;
- recycling materials and energy;
- conservation of resources;
- waste treatment etc.

These tactics must be combined with certain design and process engineering techniques, such as:

150

- providing a considerable excess of the least hazardous agent,
- minimizing dwell times,
- recirculation of materials and energy via closed loops,
- concurrent reactions and product separation,
- introduction of heterogeneous systems,
- adaptive processes and apparatuses,
- increasing throughputs,
- multifunctional environmental facilities etc.

For a cleaner economy to be affordable an environmental market must be established and market mechanisms set in motion between all its interacting operators. Integration is necessary that would link researchers and developers of environmentally high technologies to designers of equipment, to manufacturers, to users. Professionals must be trained and further educated in the fields of industrial ecology and environmental management. Qualified consulting and assessing bodies are needed that would be capable of certifying environmental products, properly performing scientific, engineering and legal assessment, and winning public trust at the environmental market. Legislation is necessary that would give incentives to managers and entrepreneurs promoting cleaner production, ensure benefits to companies upgrading their production facilities to make them more environment-friendly, and stimulate development of an environmental market focused on high technologies, equipment, labour and services and having all proper attributes such as competition, arbitration courts, commercial practices etc.

The above paragraph is closely related to the idea of restructuring in the area of material production based on:

- developing a socially oriented market economy that would guarantee a proper life standard for the population;
- cleaner production, minimizing environmental loads, material conservation, adoption of new types of activity grounded on environmentally safe technologies;
- making a more balanced economy by shifting from production of means of production to consumer goods, and
- environmental impact assessment and auditing for all economic projects.

The macroeconomic transformations rely on changes in the structure of production and consumption, mainly in industry. This necessitates:

- a more pronounced social orientation of industry to increase the relative importance of light and food-processing industries;
- an effective combination of industry branches keeping abreast of international requirements and meeting domestic needs;
- setting limits to raw material and semi-finished product industries;
- step by step reduction of exports from primary and other material- and energy-intensive industries;
- increasing outputs of high-added-value products to facilitate effective utilization of domestic resources, and restructuring the production environment via introduction of recent scientific accomplishments;

- conservation of energy and other resources;
- implementation of waste-free and environment-friendly technologies, application of optimised power sources, waste treatment and utilization.

All over the world, the environmental market is replacing punitive methods of environmental management and those environment protection agencies that are not capable of coordinating and managing cleaner economy projects.

Throughout the world, the taxation and payments for resources and emissions are devised in such a manner as to make it more profitable for the manufacturer to resolve the environmental issues in-plant, rather than to shift them to the consumer area. A combination of sanctions with economic incentives for cleaner production will make the latter not a recipient from, but rather a donor to the government budget.

Yet even the predominance of a cleaner economy policy will not immediately guarantee the survival of the population in a deep economic crisis like the current one. The anthropogenic damage already caused to nature may prove too severe and not lend itself to repair within the life span of one or even more generations.

An analysis of the man-production-environment system reveals that for the survival of human beings, the CP concept must be complemented by two further lines of action, namely:

- adaptation of human body to life in adverse conditions, and
- utilization of life support systems.

The former approach implies development and implementation of biomedical and nonconventional methods for prevention of ailments, adaptation and rehabilitation based on recent scientific findings and combined efforts of scientists, engineers, educators and managers under a degrading environment in Ukraine. These are the major lines to be pursued:

- setting up a tailored system of environmental education and training for the population in environmentally damaged areas, relying on the existing environmental education network and the media;
- research and development of adaptation promoters, immunogens and detoxicants, mostly of natural origin, processes and equipment for their manufacture and application practices;
- launching industrial production of adaptation promoters, immunogens and detoxicants, mainly from Ukraine's domestic starting materials; and
- research and development of existing and new non-medicinal methods of health building and adaptation to anthropogenic loads, including ways to reduce immune reactivity of and risks to people subjected to adverse working environments, residing in heavily polluted areas or dealing with ionizing radiation and other negative factors at work.

The latter concept implies providing local life support systems for unfriendly environments. Ukrainian scientists and engineers have already developed a variety of processes for potable water treatment by adsorption, electrochemical oxidation, electrocoagulation, electro-coprecipitation, electrodialysis, electrofloatation, floatation, membrane techniques etc. Each family must be given small units for water purification, air cleaning and removal of hazardous substances from the food as soon as possible, for it may take decades to introduce cleaner production on a national scale. Here, we should follow the example of Western business people who bring with them to the Ukraine devices enabling a safe existence in this unfriendly environment.

More specifically, environment professionals in Dnepropetrovsk have offered a number of local CP projects. One of them is concerned with the treatment of ash from the local steam power plant. According to Canadian experts, 32 elements may be recovered from the ash, in addition to the residual coal, making the business of ash treatment highly profitable.

There have been projects to produce building materials from the fly ash collected directly in precipitation filters. Moreover, this material attracted international entrepreneurs who wanted to export it to Spain, most probably for the purpose of extracting some rare earth metals. It is regrettable that no local business people showed interest in the idea, especially when in the Dnepropetrovsk area there are defence industry giants, such as the Yuzhnyi Engineering Plant, with their expertise in high technologies, including recovery of valuable metals and fabrication of appropriate sorbents and equipment.

The steam power plant is to become another site for an exciting project enabling a 2-fold reduction in the degree of flue gas cleaning while cutting the electric power consumption by a factor of 2 to 3. The new process that applies pulsed voltage to the precipitation filters has been successfully introduced at several other plants in the Ukraine.

These and other projects were included in the draft programme of cleaner production for the Dnieper region. Each item in the programme is backed with engineering and economic analyses. For many of the projects, international partners and prospective investors are sought who may gain profit by cleaning our environment.

The regional programme of adaptation and rehabilitation of the population is being applied for the first time in the Ukraine as for near Dniepr river region, and in a kind of the concept is offered for any of a technogenous intense region. A feature of the programme is the integration of efforts and joint activity in science, engineering, training and management aimed at addressing problems of preventive maintenance, adaptation and rehabilitation of the population in worsened ecological conditions in the Ukraine.

The main sections of the programme:

1. Setting up of a system of environmental education and training of the population, and direct work in environmentally severely damaged regions using a network of ecological training and mass media,

2. System engineering of diagnostic, adaptation, the improvement of health standards and rehabilitation of people whose health has been detrimentally affected by environmental pollution.

3. Scientific research on ordering existing and developing new non-medicinal methods of health improvement and adaptation to technogenic pressures, in particular, development of ways of increasing immunity and reducing risks for people exposed to harmful industrial factors,

4. Scientific research on the development of adaption promoters, immunogens, and other medicines which raise resistance to the effects of harmful substances (predominantly of natural origin), application methods, manufacturing technologies and equipment. Survey of natural raw sources of adaptogenous, immunogenous and development of industrial technologies and creation of its manufactures.

6. Development and introduction of new food products on the basis use natural byoaddings with adaptogenous and immunogenous properties, ensuring preventive maintenance of diseases ecological ethyology.

7. Development and organisation of manufacture of an ecological engineering and surviving means.

In order to optimise the natural environment, an ecological hydrochemical valuation of the objects in question, using normative limiting parameters of environmental quality and methods of ecological optimisation, is required. Forecasting and regulation of optimal processes, evaluation of efficiency of use of mineral resources, control of natureusing, regional economy and ecological monitoring with technical, programme and information maintenance are all necessary features. And, finally, we have to ensure the control and management of the quality of the natural environment based on the science of the biosphere and its evolution.

So we hit on the necessity of using systematic ecology based on concepts of system analysis. To this end we require databases on the physical-chemical methods of effect on systems and of methods of optimisation and ecologisation. In order to ensure a comprehensive approach it will be necessary to use criteria which describe the optimisation of ecological engineering and provide a quantitative definition of the intensity, efficiency, flexibility and degree to which clean production is implemented.

Thus, it is necessary to develop a system which provides a quantitative evaluation of the ecologisation of industrial objects and of the extent to which ecosystems have been destroyed, in order to ensure acceptance of justified decisions directed at the ecologisation of industrial and agricultural manufacture and at rehabilitation of areas polluted by toxic substances.

An important problem is the integration of processes of accepting decisions made in the field of the environment and development, and the improvement of transfer systems and analytical methods. Comprehensive measures which evaluate the consequences of the decisions in economic, social and ecological terms are necessary, not only on the level of the individual projects, but also on the level of policy and programmes. The analysis should include an evaluation of costs, profits and risks.

Using a system-structural analysis we deal first of all with the ecologisation and optimisation of chemical and metallurgical plants and, in particular, of their equipment. We also deal with research on cleaner production technology and equipment, and analyse the interaction of industry and the environment. This enables us to determine, on the basis of system-structural analysis, which improvements in technological processes would ensure the greatest reduction in environmental impact, and, hence, to determine a strategy and tactics for cleaner production. Some specific methods of ecologisation that can be employed along with traditional methods in any field of engineering (closed-loop structure and multifunctionality of equipment, intensification) are as follows:

- minimizing treatment time and the excess of one of the reagents, resulting most frequently in increased selectivity and in a reduction of by-products;
- recuperation, closed cycle of substantial and energetic fluxes, had shown in "idealization" of the synthesis regimes and in a significant reduction of the speed of the secondary reactions rate;
- combining the synthesis and separation and heterogenization, resulting in a significant reduction in by-product formation by carrying the target products outside the reaction zone the moment they are formed;

154

- adaptability of methods and equipment, ensuring the reliability of the technical system through the "intrinsic" reserves of the installation, thus minimizing the risk of polluted "volley" ejections from the installation.

As mentioned above, experience from all over the world testifies that the main trends in ecologisation are the following: development of low- and non-waste technological processes and equipment, and rendering industrial and domestic wastes harmless.

From a system analysis point of view, the problem of interaction between humans and the environment in accordance with development of a industry more and more significant role begin to play feedback, i.e. influence of a environment to development of manufacture. In the opinion of many experts in the field of industrial ecology it is necessary to aim not at complete clearing or waste salvaging, but to admit the existence of industrial wastes and aim for non-waste production or, at least, for cleaner technology. It is necessary to evaluate all technologies in terms of their potential ecological danger – in terms of the quantity and quality of the waste produced. Parameters for the quantity and quality of pollution caused by gases emitted, by waste waters and by solid waste are the most objective criteria indicating the imperfections of technology.

For example, M. Slavin proposes [1] using economical effect E of realization technical arrangements directed to environment protection, as the quantitative evaluation mentioned above. This can be calculated using the formula:

$$E = \sum_{i=1}^{i=10} E_i, \quad (1)$$

where:

E_1- value of supplementary products or energy obtained as a result of utilization and using waste and lateral-side products;

E_2- costs of transportation, production and consumption waste to the centres of destroy transformation or utilization;

E_3- cost of restoring land and reservoir structures or on the negative consequences when they cannot be put to rational use;

E_4- costs incurred from environmental damage to or destruction of animals and plants, rendering their use inadmissible or inefficient;

E_5- cost of medical treatment, diagnosis, preservation and nursing of people suffering from poisoning or intoxication, caused by unfavourable environmental influences;

E_6- reduction of money spent by the state on temporary disability due to illnesses caused by unfavourable environmental influences;

E_7- liquidation negative economic consequences when production costs are not covered due to the loss of working ability of persons being in zone of the productional influence on the environment;

E_8- cost of restoring building facades, gardens, parks and other places intended for workers' leisure time and of restoring buildings and memorials in the charge of the state;

E_9- cost of planned and unplanned monitoring of environmental conditions by sanitary epidemic stations or other sanitary and hygienic services;

E_{10}- liquidation negative economic consequences when production costs are not covered due to temporary stops in production resulting from violation of recommended sanitary and hygienic standards.

The ecological influence criterion (e.i.c.) which makes it possible to measure standards of production technology with regard to interaction with the environment, is proposed by V.Anikeev and co-authors [2]. They suppose that " the most representative and useful ecological criterion is the ecological influence criterion k (e.i.c.)", which is defined by the formula:

$$k = W_t / W_r = W_t / (W_t + W_c), (2)$$

where:

W_t- theoretical influence necessary for production;

W_r- real influence;

W_c- influence determined of concrete production.

The maximum significance of e.i.c. equals 1 is determined by the condition $W_t = W_r$, that is when real influence corresponds to the theoretically necessary level determined by laws of substance and energy. A lower e.i.c. requires more technical solutions as regards environmental impact. And when k-- 0, production disregards the demands of environmental protection.

The following indexes [3] for measuring the ecological efficiency of technological processes are also proposed:

- ecological measure (L) - quantity of damage to the environment (P), divided by the quantity of useful production or services obtained in this process (Q):

$$L = P/Q, (3)$$

- resource capacity of the process (N) - debit of energy, water, air, ground and other natural resources (R) to the useful production or services obtained in this process:

$N = R/Q , (4)$

- ecological index of the object (E) defined by the formula:

$E = (Q - P)/ R, (5)$

where (Q - P) is the useful effect.

All sizes of Q, P, R are getting in natural significance. It appears to the authors that the ecological index of the object (E) describes the degree of the object closing in relation to nature. When E = 0, the nature potential destroys without any useful effect, and when E = 1, there is not non-utilization remainders of the matter or energy.

Thus, it is necessary to develop a system which provides a quantitative evaluation of the ecologisation of industrial objects and of the extent to which ecosystems have been destroyed, in order to ensure acceptance of justified decisions directed at the ecologisation of industrial and agricultural manufacture and at rehabilitation of areas polluted by toxic substances.

According to J. Wotte [5] an accurate way of evaluating technological processes in order to develop and to introduce LOW AND NON-WASTE TECHNOLOGY (LNWP), would, theoretically, be by cost-benefit-analysis. Not only technological and economic influences should be taken into consideration in this connection but also ecological and social aspects. The question of financial evaluation of environmental impacts and similar effects, however, has not yet been solved at an international level. Therefore, other ways must be sought for evaluating technological processes.

A new method was tested [5] and completed especially for industrialization purposes in developing countries. The method developed may be used:

- to compare several technological processes (at least two) for the production of a defined product;
- to evaluate the processes from the standpoint of LNTW;
- to make a clear choice as to which process variant should be preferred on account of its technological, economic, ecological and social characteristics and their relevance to both national economic conditions and as regional or local prerequisites

The developed method is based on selected characteristics which are grouped to form a minimum programme, characterizing the technological process and, if necessary, the intended main product from the standpoint of low-and non-waste technology (LNWT).

The steps of the selection process represent a combination of evaluating as well as weighting the given characteristics.

For evaluating the characteristics against the background of a global level analysis there are estimated for each characteristic:

-the zero value, K_o (accepted worst value) and
-the best Value, K_b (best possible value).
The evaluation is calculated using the equation:

$$PW = \frac{K_o - K}{K_o - K_b}, \quad (7)$$

where:

 K – real value of the given characteristic. This delivers values for all characteristics for all technological variants under discussion within a given industrialization project.

For weighting the characteristics a group of experts is formed consisting of technologists, economists, ecologists and land planning specialists. All members of the group should be skilled and experienced specialists in their profession or their fields of activity, and their private interests should not in any way be connected with the interests of the enterprise ordering the selection procedure.

After the group of experts have been thoroughly briefed on the advantages and disadvantages of the technological variants under discussion as well as on local conditions and the national economy as a whole, each expert will be given a questionnaire, in which the characteristics are grouped in pairs for comparison.

Each expert separately marks in each pair of characteristics, which of the two he or she considers to be of greater importance. On the basis of the choices made in this paired comparison a personal preference table is compiled for each expert.

The judgments shown in all the experts' choices are tested using well-known methods of statistics. In order to optimise the natural environment, an ecological hydrochemical valuation of the objects in question, using normative limiting parameters of environmental quality and methods of ecological optimisation, is required. Forecasting and regulation of optimal processes, evaluation of efficiency of use of mineral resources, control of nature use, regional economy and ecological monitoring with technical, programme and information maintenance are all necessary features. And, finally, we have to ensure the control and management of the quality of the natural environment based on the science of the biosphere and its evolution.

The fact that, ecologically speaking, all kinds of natural resources are interrelated means that it is crucial to perform a system analysis of their present condition, including a complex forecast of future use and a definition of the extent to which natural resources may be exploited. In order to perform a comprehensive quantitative evaluation of a system's sustainability it is helpful to use the integrated parameter of sustainability, that may be to define as multiplicational coefficient including the main technical characterizations of technological process connected with resulting measure of production purity such as: flexibility of production; intensity of methods; efficiency of technology, etc., and also parameters, that reflect, accordingly, social and economic efficiency of transformations, determined by expert way. Integrated parameter S (parameter of sustainability), determined as product:

$$S = aE \times bR \times cK, \ (8)$$

where:

a, b, c - "weights" contributions of relevant parameters,

R, K - parameters reflecting, respectively, social and economic efficiency of transformations, determined by expert way,

E- parameter of a system's ecologisation, which it is useful to define proceeding from following reasons.

In our opinion, the degree to which production improves with regard to its influence on the environment may be estimated as a multiplicational ecologisational coefficient:

$$J = \prod_{i=1}^{k} J_i , \ (9)$$

where:

J_i are the main characteristics of a technological process connected with resulting measure of production purity such as:

J_1 - flexibility of production;

J_2 - intensity of the technique;

J_3 - efficiency of the technology, etc.

Any of these measures may be defined similarly to the efficiency of any technological object as:

$$J_i = \frac{P_{1i} - P_{2i}}{P_{maxi} - P_{2i}} , (10)$$

where P_{1i}, P_{2i}, P_{maxi} -are respectively the final, initial and maximum measures.

The value P_i may be both a common index characterizing quantitatively one or the other property of the system that is connected with the corresponding measure (for example, the dust collecting efficiency, extraction ratio, etc...), and a complex integral index, which counts several characteristics of the object simultaneously.

In the latter case it is supposed to use additive indices such as:

$$Pi = \mathbf{S} \, K_j \, P_{ij}, \quad (11)$$

$j=1$

where K_j is significance of the j-th indice (estimated expertly and changing within the limits of O to 1).

It is also necessary that the multiplicational ecologisational coefficient J for comparison of the different technological installations contains the identical typesetting J_i.

It is only possible to connect these databases, approaches, concepts, etc. in a joint system by applying technical theories and methods and employing the laws of the development of technical systems, using psychological features of creative activities and modern methods of technical creative activity in solving ecologisation problems (such as the method of trial and error, brain storming, morphological analysis, synechtics, theory and algorithm of inventory tasks solution, decision making in risk conditions, elements of the theory of usefulness, means of support of decisions based on information, means of the analysis and modelling of ecological situations; methods of analysing conditions, forecast of development, modelling of the decisions on waste problems; expert valuations of proposals on handling waste problems.

It is possible to solve the complex problem of integrating economic, biological and human systems via collaboration., if collaboration is pursued among the engineering, information, mathematical modelling, and ecological communities.

3. Tiers 3-8. Engineering techniques and methods for Cleaner Production

Engineering techniques and methods for cleaner production seems somewhat limited and lacking in diversity. The reason for this may lie in the unwillingness of some users to disclose their know-how, or simply in the absence of new approaches. At the same time there are many other effective means of improving product cleanliness. We use, for example, the following cleaner production tools and methods:

Flexible synthesis systems and adaptive equipment

Mtehods such as process engineering for high throughput to cut processing time and reduce by-products and wastes, and industrial symbiosis as a basis for management of secondary materials and energy, can be useful tools. They lead to a minimization of processing time and less surplus toxic reagent, all resulting in an increase of selectivity and reduction in the volume of by-products formed. Further processes that may be applied are:

- Synthesis and separation in an aerosol to increase intraparticle pressure and reaction rate.
- Self-excited oscillation of reacting phase flows at frequencies and amplitudes matching those at the rate-limiting tiers of the system.
- Recirculating flow of the least hazardous agent taken in excess over its stoichiometric value.
- Isolation (close-looping in structure) of flows of substance and energy by recirculating, resulting in "idealization" of modes of synthesis and significant reduction of the speed of by- processes.

- Separative reactions organizing (synthesis and dividing processes organising in the same place and at the same time), reducing formation of by-products by removal of a target product from a reactionary zone at the moment of its formation.
- Controlled heterogenization of the contacting phases for softer conditions and improved selectivity.

The above methods lead to a higher flexibility and adaptability of technology and equipment, ensuring the technical system works reliably by making use of "internal" reserves (flexibility) of installation which reduces the risk of pollution by harmful substances or reception of a sub-standard product.

The characteristics of some new concepts, principles and methods of ecologisation are slowling becoming known. The system analysis of chemical engineering works allowed us to determine the main principles and methods used in securing ecologisation of their processes. These were based on a number of different concepts, some of which were of a general technical character (recuperation, waste reclamation and resource economy), whereas others are of particular significance to chemical production, such as, for instance, the idea of providing wasteless operation not by reclamation or resource saving, but by promoting selectivity, that is the yield rate of the target product. Basically, this concept is aimed not at trying to eliminate waste, but at running the process so as to minimise the quantity of waste produced.

Along with the above stated principles of complexity and system character, it is important to mention the notion of flexibility, which is of great importance to modern engineering principles. Flexibility in chemical engineering incarnating its indissoluble unity with the technique should involve a quantitative index, which reflects the ability of the technology and equipment to function within a wide range of changements of outer and inner parameter settings with assigned values of selectivity, and consequently, the formation of by-products. At the same time, many of these methods aim at realizing the principles of "repeated use of resources and energy" and "maximum selectivity of the synthesis and separation", the meaning of which is self-explanatory.

Some tools and methods for CP are described below.

a) Geterogenization

The best results may be achieved when geterogenization is used as a part of Reactive Separation Processes (RSP) ideology (see below). We discuss below a number of possible mechanisms of chemical reactions improving with increasing of their selectivity when a mass transfer process combines with a chemical process in particularities at bubbles mode of phases interaction.

Such an approach enables a greater concentration of reagents in the reactionary area, modificated a system in heterogenous one and removes the products of reaction in the phase where reaction does not go, which reduces the chance of by-products forming to the account parallel running reactions with products of main reactions.

RSPs are in principle distinguishable from chemosorption processes in that reaction products promote an increase in the velocity of reversible process in accordance with the Le Chatelier principle, but for inconvertible - in consequence of the law of action of masses, since in the reactionary mass at the product tap to reactions increases reagents concentration.

Increasing RSP velocity promotes a faster change in surface tension on the border among the phases and their density which causes a reinforcement of surface turbulence which, in turn, accelerates a mass-transfer process - a reaction product evacuating from the liquid phase in gas that, in turn, makes the chemical process in the liquid more intensive. On hand strong mutual influence of reactionary and mass-transfer processes. This phenomenon can be not described by dependencies, tinned by joint deciding the equations of diffusion and kinetics.

An important feature of RSPs is that reactions products departing on the measure of their formation to the account of mass-transfer process imposition, it is possible to do practically constant, (or even enlarging) during concentration of source materials in the liquid phase. At the constancy of concentrations of reagents corresponding concentration multipliers in the kinetic equation possible to comprise of the velocity constant, herewith apparent order to reactions decreases.

The imposition of bubbling separating process on chemical processes can cause an increase in the observed velocity of the process, a variety of diffusion effects, in accordance with formation an interphase surface, breakup of) continuity under gas bubbling in the liquid and forming areas of increased pressure, phase transition at the interleaving of condensations and evaporations on each contact stage, introduction to the system of a quite numbers of energy with the gas flow from outside.

When a homogeneous system is transferred to a heterogeneous one, for instance, under bubbling, as is generally known, pulsation of pressure and velocities occurs, causing effects, similar effects, to appear at the influence to the ultrasonic fluctuation system.

The appearance and collapse of bubbles under steam (gas) bubbling in the liquid is related to in than-that to the phenomena an cavitations, under which in liquids appear pulsing bubbles, pervaded by steam (gas) or their mixture (at desorption by the inert gas).

Since pulsation frequencies in a bubbling zone $(1...10)^6$ c^{-1}) on much orders frequencies less than own molecule oscillations, probably, chemical change must not occur in the system in consequence of resonance phenomena, and changing the predominant frequencies in the system in the specified range will have little effect on the dynamics of chemical conversion. However, appear acoustic currents, causing an intensive mixing of phases and accelerating in several times heat-mass-transfer processes, since action of acoustic flows turns out to be vastly more efficient than hydrodynamic ones, because of the smaller boundary layer thickness. In the fluid phase herewith appear effects, similar cavitations ones, causing the bubbles to grow and to its departure by acoustic currents.

Formation and complex bubble moving (for example, the effect of "dancing" bubbles), change their once-measures, slamming, coalescence generate pulses of compression in liquids and can cause a local warming of the phase. So, under the adiabatic compression of cavitations the temperatur of the bubble can reach 10^4 K. Increasing temperature promotes the transfer of molecules on the border and in bubbles in an agitated condition and fission them on radicals (relationship breakup and forming of the radicals, according to different sources, occurs at 350-1000^{0}C), which can recombine and interact on the known mechanism. The velocity of formation and spending the radicals, probably, will render herewith-significant influence on the general integral velocity of process.

Besides, under adiabatic compression of the bubbles or expansion of their walls, a double electrical layer and electrical charges appear, and magnetic fields appear in the gas-liquid layer. At the same time it is known that even relatively weak electrical and magnetic fields can have an essential influence on the velocity of chemical reactions in the free-radical mechanism as a result of a certain order in the location of free radicals in the reactionary mass and changing a fate of the triplet couples under the influence of an external magnetic field.

It is not expelled that for many chemical reactions with free-radical mechanism transfer from a homogeneous to a heterogeneous system by gas bubbling and organization, in particular, reactionary-desorption process in consequence of complex above effects promotes to fission the molecules on radicals, increasing their concentrations in the reaction area.

Therefore, it is difficult to offer a single explanation of reasons of improving and selectivity increase for many chemical reactions at the imposition of separative mass-transfer process, in particular, running in the bubbling mode. This question requires further detailed research. However, investigations conducted by the author of this study and experience in using a series of new RSPs in industry make it possible to do highly important in the practical attitude a findings on practicability of organizations one or another combined process and, above all, for furthering clean production and preventing pollution.

The author confirmed the advantages of heterogeneous over homogeneous systems for a large number of chemical reactions and many new technologies. A profound difference in the observed velocities of chemical reactions indicates the advantages of bubble mode and demonstrates the effects of speedup, a distinctive feature of the bubbling process, as opposed to film mode of interaction desorptive agent with the fluid reactionary mass.

b) Parallel Reactive Separation Processes (RSP)

Parallel reactive separation processes (RSP) are used as clean reaction technologies for increasing purity of production, waste reduction and for pollution prevention.

The reactive zone and the distillation zone are usually individual zones in similar units. However, it can be extremely beneficial to combine these zones in one unit. If this is done, not only will the reaction heat in the reactive-separation zone cause additional mass transfer between vapour and liquid phases but the chemical process will also be speeded up. Running a chemical reaction in the same place and at the same time as a physical process of separation in the resultant reaction zone is an effective way of improving chemical processing rates. The reason for this is that, by removing reaction products as they form, reversible processes are promoted in accordance with the Le Chatelier principle, and irreversible processes are influenced, in accordance with the law of mass action, because reagent concentrations in the reaction zone are increased as the products are removed. With increasing rates of reaction and mass transfer, the interface tension and phase densities change more rapidly, resulting in more vigorous surface turbulence. This promotes mass transfer, namely the removal of products from the liquid into the gas phase, which, in its turn, increases the rate of reaction in the liquid. Laboratory and industrial research revealed that the reaction-separation mode is well-suited for acylation, amidation, amination, condensation, cyclization, dehydration, etherification, halogenation, hydrolysis, oxidation and other chemical reactions.

So, undertaking a chemical reaction together and simultaneously with the physical reactive separation process (RSP) is not only an efficient means of intensifying technical processes, but also of reducing the speed at which by-products are formed and reducing, or frequently preventing, the formation of wastes and pollution.

162

References

Slavin M. (1979) Social-hygienic factors of new technique // "The questions of economics".-Moscow: 1979.-N5, pp.58-66

Anikeev W.A., Kopp I.Z. and Skalkin F.W. (1982) Technological aspects of surrounding defense.-Leningrad: Gidrometeoizdat. 254 p.

Environmental protection. The models of social-economical prognosis.- Moscow: Economica, 1982.-224 p.

Wotte, Joris (1994) Selection of Technological Processes from the Standpoint of LOW AND NON-WASTE TECHNOLOGY for Industrialization Projects. In: Journal of EIA Vol.3, No.2 November 1994, pp.9-14.

CHAPTER 17

Results of the EcoLinks Project devoted to cleaner metallurgical production in the Donbas Region - Options for cleaner recycling of nickel bearing wastes at the Konstantinovka plant "VtorMet"

Prof. V. M. Sokolov[1], J.J. Duplessis[2], C.G. Rein[3], E.A. Pearson[3], Z.V. Kuchma[4]

[1] PTIMA NASU, 34/1 Vernadsky Ave., Kiev-142, 03680
Ukraine
[2] 715 Freeman Lake Rd. Elizabethtown, KY 42701
USA
ESS, Inc., 401 Wampanoag Trail, Suite 400, East Providence, RI 02915.
USA

[4] URDIISPOG, Uzhnoe Highway, 1, Zaporozhye, 69032
Ukraine

Summary

This chapter deals with the investigation into the introduction of cleaner metallurgical production in the Donbas region. Its aim is to study the situation at a local metallurgical facility - Konstantinovka plant "VtorMet" - and to verify measures for its improvement. This work was carried out within the framework of the International Program EcoLinks and financed by USAID. The problems considered are common to many metallurgical facilities located in the Newly Independent States. Therefore the results of the study have good potential for transferability. The recycling of Ni-bearing wastes and scraps including discarded nickel-iron accumulator batteries was improved from an environmental and energy consumption perspective. Removing the plastic components of the batteries from a charge and arranging automatic control of graphite electrodes operating in an electric arc furnace has brought about sound environmental improvements and considerable energy and graphite material savings. An action plan for further technology improvements was worked out, including identifying appropriate and cost-effective air pollution control equipment.

Introduction

The Konstantinovka plant "VtorMet" has recently developed a technology for manufacturing ferronickel from Ni - bearing waste in an electric arc furnace (EAF). Discarded Ni - Fe industrial accumulator batteries are a basic element of the wastes in the furnace charge. This type of waste is generated in significant quantities by the mining industry in the Donbas region where the Konstantinovka plant is located. This technology is new for the plant and developments along these lines became necessary after the Ukraine became independent. The plant produces nearly 5000 tons of ferronickel annually using this technology. This recycling leads to significant emissions of dust and gaseous components hazardous to human beings and to the environment. Thus, the plant is greatly interested in reducing the volume of emissions by melting technology improvements as well as by setting up a gas purification system. The dust that will be caught by the system that is envisioned contains a large amount of nickel. It will be returned into production for remelting, thereby increasing the process yield. The plant has some other problems requiring improvement.

163

W. Leal Filho and I. Butorina (eds.),
Approaches to Handling Environmental Problems in the Mining and Metallurgical Regions, 163–174.
© 2003 Kluwer Academic Publishers. Printed in the Netherlands.

The most important of these are the relatively high consumption of consumables, electric power and graphite, and the lack of proper slag utilization.

These important issues were addressed by the joint Ukrainian-American project "Environmental friendly recycling of Ni - bearing wastes at the Konstantinovka plant "VtorMet". This project was conducted between May 2001 and April 2002 within the framework of the environmental programme EcoLinks financed by the USAID. The Company "VtorMet" took part as project leader to initiate the work. Environmental Science Services, Inc. (ESS) assisted as a partner, and the leader's efforts were facilitated by several Ukrainian experts. ESS is an employee-owned environmental consulting and engineering company with offices located in Wellesley, Massachusetts and East Providence, Rhode Island. The Ukrainian Research and Development Institute of Industrial and Sanitary Purification of Off - Gasses (URDIISPOG) carried out some environmental measurements and examinations as an associate in the project.

Project Purpose

The overall goal of the project was to identify and assess the environmental problems resulting from recycling discarded Ni - Fe industrial batteries, and to develop and test environmentally sound, friendly and efficient measures that would provide a system to reduce air pollution and lead to improvements in the manufacturing process indices, such as the consumable usage and energy consumption. It was also hoped that a use would be found for the slag products. The individual objectives of the project originated from the above overall goal. They included improving the basic elements of the manufacturing process at "VtorMet" to promote a more environment-friendly and cost-efficient recycling process.

The major elements of the recycling process had to be examined. This included examining the content of the discharged gases, the methods of preliminary preparation of raw materials, electric energy consumption, the amounts of consumables consumed, and the slag utilization process. Improvements were necessary which would reduce the hazardous constituents in the gas emissions that pollute the plant and neighbouring areas while at the same time leading to more efficient recycling process indices.

Air pollution emission measurements were needed for the design of a modern off gas purification system that would lower emissions of the most harmful constituents of the gas emissions to the level required by Ukrainian environmental laws.

A further target of the project was to define effective measures for reducing consumption of electricity and consumables and creating a proper system of slag utilization. The feasibility of introducing a system of controlling electric power consumption during melting at the VtorMet plant was also to be assessed.

Identifying and evaluating the environmental and technological problems

3.1. EAF FURNACE OPERATIONS

The investigated EAF is located near one end of a large building with the approximate dimensions: 61 m long by 21 m wide by 15 m high. Along the centre of the roof and running the length of the building, are vents/windows that create a chimney effect draft to allow heat and smoke to escape by natural ventilation. An array of rectangular exhaust hoods are positioned on the EAF cover for local collection of smoke and fumes, which are then ducted outside the building through a ground-mounted 75 kW fan and up a 22-meter tall unlined steel stack with an inside diameter of 0.98 metres.

Though relatively small in size, this 5-ton EAF is designed and operated in a manner similar to those in the United States (US) and elsewhere. Scrap batteries in large metal holding containers (hoppers) brought into the building from the scrap yard are loaded into the EAF hearth by an overhead clam-shell crane bucket in a series of batch charges. During charging the EAF hearth is moved horizontally out from under the EAF roof and exhaust hoods, and the clam-shell drops the charge into the hearth, generating a cloud of dust and dirt that drifts up to the roof. After charging, the hearth is moved back under the roof and exhaust hoods, and then three electrodes are lowered into the charge. The roof does not fit tightly on the EAF hearth, but is positioned about a metre above it. As the charge melts, a great deal of smoke and fumes are generated, some of which are captured by the roof hoods whilst some escape to the room air and drift to the roof vents of the building.

After a melt is complete, the hearth is first tipped to pour the slag off the top into a large ladle, and then the bottom tap hole is opened and the hearth is tipped to pour the molten metal into another ladle. In pouring both the slag and the molten metal, clouds of smoke and fumes are generated, all of which rise to the roof of the building. The slag ladle is carried by a second hook crane to the slag pile in the furnace building and dumped. The molten metal is carried by the hook crane to the place where the metal is poured into ingot moulds. The slag dumping and ingot pouring generate smoke and fumes which also drift to the building's roof vents.

The EAF's roof exhaust hood ducts manifold together and interface with the duct that penetrates the building wall via two matching flange faces. The face attached to the hood duct rotates against the other face when the EAF hearth is tipped, thereby opening up the connection and yielding less suction at the roof. As a result, the roof hoods capture little or no smoke/fumes during slag pouring and metal tapping. From outside the building, smoke and fumes are visible coming from the building vents, and the plume from the hood exhaust stack is very dense (100% opacity) and of a grey/yellow colour.

These observations confirmed that the process of ferronickel production from secondary raw material - discarded nickel-iron batteries and scrap containing iron – is accompanied by a high level of emission of gaseous and solid substances to the atmosphere. It was, therefore, necessary to work out a package of measures to solve the problem of reducing the emission of harmful substances. This package would include improving the remelting technology and the process of raw material preparation, and developing an efficient and reliable system of pollutant removal. First of all, it was expedient to examine thoroughly the characteristics of the waste gas emitted in the process of scrap remelting.

INITIAL AIR EMISSIONS TESTING

Two air emission test programmes were conducted during this project: one in May 2001 and one in February 2002. During the first test programme, measurements were made in the exhaust stack during normal melting of accumulator batteries. Measurements were made for the following parameters: volumetric flow rate; particulate matter (dust); nitrogen dioxide (NO2); hydrogen chloride (HCl); sulphur dioxide (SO2); carbon monoxide (CO); hydrogen (H2); low and high boiling organic compounds.

Dust samples were analysed for their iron, nickel, potassium and sodium content, and also underwent a limited particle size distribution analysis. Dust samples were also analysed for various physical characteristics including density, angle of repose, and resistivity. Air emission samples were collected and analysed using standard procedures and equipment normally employed in the Ukraine and other eastern European countries. The results of these first tests are summarized in Table 1.

Using these test data, ground-level concentrations of the most harmful substances were estimated according to the EOL programme recommend by the Ministry of Environment and Nature Resources of the Ukraine. The results showed that the concentration on the ground of benzo(alpha)pyrene in the area 300m from the stack exceeded the maximum permitted concentration.

To identify the changes required to correct this unacceptable situation, a bench-scale trial was carried out. It consisted of heating in a tube electric resistance furnace a charge of material, this being the scrap from a single nickel-iron accumulator. The temperature was raised to 1200°C. It was observed that after the temperature had risen to 1800°C, a dense white smoke began to be given off. When the temperature was raised further this was converted into a yellow-green volatile matter smelling like chlorine. This smoke was not caught by the aluminum oxide and water trap. After reaching the temperature of 3800°C the smoke had the specific smell of a mixture of high-boiling hydrocarbons. The product of its condensate did not dissolve in water and alkaline solutions. A carbon substance in porous form was left in the charge after it had been heated up to 600°C. It became obvious that after the charge had been heated up to 1800°C the thermal disintegration of a PVC separator from the accumulator began. This was accompanied by the extraction of stable aromatics and polycyclic compounds. The process of hydrocarbon chains chlorination at double bond took place simultaneously. Chlorine-organic substances were formed. The passing of the above gaseous products through the liquid slag during the melting campaign lead to the formation of hydrogen chloride, chlorine, chlorine-organic products, cyclic and aromatic compounds.

Table 1. Comparison of Air Emission Test Data VtorMet Electric Arc Furnace Konstantinovka, Ukraine

Test Date	Melt Condition	Flow rate[b]		Dust Concentration[a]		Dust Emission Rate[c]		Nickel (Ni)		Iron (Fe)		VOC[d]		SVOC[e]	
		(Nm3/hr)	(scfm)	(g/Nm3)	(gr/scf)	(kg/hr)	(lb/hr)	(kg/hr)	(lb/hr)	(kg/hr)	(lb/hr)	(kg/hr)	(lb/hr)	(kg/hr)	(lb/hr)
May 2001	100% accumulato r batteries	44 136	27 877	0,327	0,133	14,4	31,8	0,210	0,460	1,76	3,89	2,6	5,7	13,8	30,4
Feb 2002	92% misc. nickel scrap 8% metal turnings	50 760	32 061	0,117	0,0475	5,9	13,0	0,028	0,061	0,070	1,54	nd	nd	1,9	4,3

a) g/Nm3 = grams per normal cubic meter at 0°C and 760 mm Hg., gr/scf = grains per standard cubic foot at 68°F and 29,92 inches Hg (20°C and 760 mm Hg). = (g/Nm3) x 7000 x 273/(453.6 x 35.31 x 293).

b) Nm3/hr = normal cubic meters per hour, scfm = standard cubic feet per minute.

c) kg/hr = (g/Nm3) x (Nm3/hr) /1000, lb/hr = kg/hr x 2.205

d) VOC = volatile organic compounds, including benzene, toluene and xylene

e) SVOC = semi-volatile organic compounds, C13 through C22, including specific target compounds, nd = none detected

The conclusion drawn from these results was the need for a preliminary step of dismantling the batteries before putting them into the furnace charge for melting. This step would reduce significantly the amount of organic emissions occurring during the melt.

ACCUMULATOR BATTERY HANDLING AND STORAGE

The procedures and equipment used by VtorMet for the storage and handling of the batteries on-site were reviewed by the ESS team together with plant personnel. Scrap batteries are received at the plant in uncovered rail cars or in uncovered trucks, which are weighed as they arrive on-site. The potassium hydroxide (KOH) electrolyte solution is drained from the batteries before shipment to the plant. Once weighed, the batteries are removed by electromagnet or by hand into uncovered furnace charge hoppers or into other rail cars for storage. The charge hoppers are then moved into the furnace building as they are required. The size of shipments to the plant range from up to 70 tons per rail car to up to 20 tons per truck.

ANALYZING TECHNOLOGICAL ASPECTS

The material balance (initial charge, auxiliary materials, and ingots produced) was calculated. The chemical composition of the metal produced and slag generated were determined, and the average consumption of the auxiliary materials used for the production of ferronickel was determined to be 12 kg of graphite electrodes, 10 kg of limestone and 5 kg of crashed glass for manufacturing one ton of this material.

The opportunities for utilization of the slag, which is generated during pyrometallurgical recycling operations were determined. The slags have been landfilled on the site of the VtorMet plant since 1932. Five types of slag piles were identified. They differ significantly from each other in outward appearance, namely colour and dimensions of bits forming the piles. A specific core sampler was used. This was simply a tube that was pushed down into a pile of sludge-like material. It filled up from the bottom with a sample representing the vertical position where the sample was taken. A second rod just big enough to fit inside the hole in the tube was used to push out the sample. The samples were taken in many places across a given pile. The sample material was then dried and mixed. A re-sample of this mixture was taken. Many samples were taken and blended to make a composite of the samples from each pile. The final sample produced was subjected to chemical analysis.

Lessons learned from the business trip to the USA

The team from VtorMet visited a number of major US companies involved in recycling Ni-bearing secondary raw materials. These were Haynes International Inc., Crucible Specialty Metals Division of Crucible Materials Corp., Veltec Corporation, J & M Industrial Inc.

Haynes International in Kokomo, Indiana, is a company known worldwide for special high nickel and nickel-based alloys. They have introduced many innovative nickel-containing alloys. The company successfully uses Ni-bearing turnings and grinding swarfs in the melting charge. These are mixed with oil-containing residues practically in the same way as the Ukrainian wastes recycled at VtorMet. The use of a dust collector (a baghouse) meets the rigid requirements of the state of Indiana Environmental Regulations.

The same method of off-gases purification during alloy steel pyrometallurgical production is used at Crucible Specialty Metals in Syracuse, New York - a leading producer of stainless, valve, high speed and tool steels for over a century. The Ukrainian team observed a melt at the company's plant. It was noted that the electric power conditions of the melt were significantly more stable than those at the "VtorMet`s" plant, since the movement of electrodes is adjusted automatically. Only once did the operator manually adjust the transformer cascade. The purpose was to obtain a long arc instead of a short one. This change became necessary after complete melting of a charge. It became obvious that this system needed to be introduced at the VtorMet plant.

The delegation's visit to Veltec Corporation in Pittsburgh, Pennsylvania led to understanding state-of-the-art principles in recycling metal beads containing slags. The common approach used for separating metal beads from slag was clarified. Jaw crushers and different types of mills (rod, ball and hammer types) are used for size reduction of the slag. Magnetic and gravity separation techniques are used for extraction of metal beads after the material has been reduced to a relatively small particle size. Several methods of agglomeration or consolidation of the separated fine wastes released by the previous operation are considered. This consideration usually results in selecting mechanical equipment designed for briquetting or pelletising. The preparation of the material for mechanical compaction requires very careful selection of auxiliary materials such as binders before a product can be produced.

An excellent opportunity for Ukrainian mining and metallurgical companies to improve their environmental facilities was revealed during a visit to J & M Industrial Inc. in Pittsburgh, a company that distributes used equipment to the above industries. The Ukrainian team were shown five baghouses of a type similar to those observed at Haynes International Inc. and Crucible Specialty Metals and applicable at the VtorMet plant. Examination of the collector showed that it had been used for only a very short time, since its conditions differed only slightly from those of new equipment. The brief examination of the other units showed that they were in the same good condition although their prices are significantly lower than prices for similar new equipment. The opportunity of purchasing used equipment seams to be very hopeful for Ukrainian casting shops, which, as a rule, are pressed for funds. For environmental reasons, however, it is not possible to use cheap secondary raw materials in order to make production more competitive.

Verification Tests and Changes Implemented - the Most Urgent Measures

AIR EMISSIONS TESTING AFTER ELIMINATION OF ORGANIC COMPOUNDS FROM THE CHARGE

The first emission tests described above were conducted at the EAF cover hood exhaust stack during normal batch melts of accumulator batteries. Some of the hood enclosure panels were removed during these tests to allow for better observation of the EAF during the melts. Removal of these panels most likely led to less smoke being collected by the hood and directed to the stack, and consequently more smoke escaping to the melt room air and exiting as fugitive emissions through the building roof vents. Thus, for the type of material melted, the emission rates indicated by the May 2001 test data are probably somewhat lower than emission rates that would occur with all the hood panels in place.

In order to document the effect of less plastic in the melt, a second emission test programme was conducted in February 2002.

The scope of the second test was similar to that of the first one. Air emission samples were collected and analysed using standard procedures and equipment normally employed in the Ukraine and other eastern European countries. A comparison of the two sets of test data is shown in Table 1.

The melt material used during the second tests did not include accumulator batteries, but consisted of 92% miscellaneous nickel scrap and 8% oily metal turnings. The data comparison shows that emissions of dust, nickel, iron, volatile (low-boiling) organics (VOC) and semi-volatile (high-boiling) organics (SVOC) were much lower during the second tests. The scrap nickel that was used in the second tests reasonably represents the dismantled accumulator batteries, and the test results demonstrate the importance of dismantling the batteries (removing the plastic) before melting. This is indicated most clearly by the fact that the dust emission rate is 2.4 times lower and the SVOC emission rate is 7.5 times lower in the second tests. The nickel emission rate is also 7.5 times lower, reflected in part by the lower nickel content of the dust samples in the second tests (0.47%) compared to the first ones (1.39 to 1.51%). The moisture content of the stack gas was approximately 0.6% by volume, reflecting a dew point of 0°C (32°F). This indicates that the moisture content in the hood exhaust is essentially that of ambient air, and therefore condensation within a control device such as a baghouse should not be a problem.

SLAG UTILIZATION

The necessity for the recovery of metal beads from the slag previously landfilled on VtorMet property and from other landfills currently in use, became obvious from the knowledge the team obtained during the visits described above. The most valuable slags are generated during the oxidation period of the melt.

The content of stainless steel beads in the slag varies from 6% to 12%. The chromium oxide content ranges from 30% to 40%. Nevertheless, blending slags from multiple sources complicates the technology of recycling. Different options of recycling slags were considered.

The best recycling option involves separation of crushed and shredded slag by the magnetic and pneumatic methods. The extracted slag obtained after separating the metal particles is optimum for use in a direct or alternating current submerged arc electric furnace for obtaining iron and chromium alloy by reduction using the wastes from aluminum production or with ferro-silicon. Moreover, the same process could be applied to non-separated slag. In this case a nickel-bearing ferro-chromium alloy is produced.

Alternatively, the separated fine stainless steel beads are an excellent charge material for melting to produce cast stainless steel alloys by the modified electroslag crucible process. The application of a processed graphite upper electrode gives temperature control of a melt over a wide range. Therefore, effective alloying and microalloying to a precise chemical objective can be accomplished. Casting at optimum temperature conditions can also be performed. The melting of the steel particles in a slag layer ensures a high purity of the produced metal. So this technology creates an opportunity for production of high performance castings.

Another major subject of study was the application of non-metallic base of slags generated after the release of the metal beads. Two types of slag were considered, namely a slag from an acid melt:
1 - 88.4% SiO_2; 2.75% Al_2O_3; 4.24% Fe_2O_3; 0.17% TiO_2; 0.19% MnO; 0.42 CaO; 2.45% MgO; 0.25% Cr_2O_3; 0.01% Ni; 0.002% Co; 0.011% Cu; 0.0283% Zn;
and from a basic melt:

- 37.6% SiO2;
- 8.38% Al2O3;
- 7.56% Fe2O3;
- 11.66% TiO2;
- 5.46% MnO;
- 8.70 CaO;
- 6.25% MgO;
- 13.0% Cr2O3;
- 0.14% Ni;
- 0.002% Co;
- 0.005% Cu;
- 0.0025% Zn.

The second slag was generated during former Soviet times and was landfilled at the VtorMet plant site. Strengthening of the banks of the river Krivoy Torets in the Konstantinovka region using this material has been a unique application of this slag. It is hoped that it will also be used for road building. The Ukraine is in the early stages of a programme for improvement of the road network that links Western and Central Europe with Russia. Building a sector of the motorway around Zhitomir was the first step in this programme where the slag was, however, provided by another supplier. Nevertheless, the slag properties required for constructing a road surface and base, complementary layers of the base, and certain building layers of a pavement should be applicable to other similar projects, so that it would be possible for VtorMet to act as a slag supplier here.

The Ukrainian Standard (GOST) 3344-83 "Slag Road Metal and Sand for Road Building" which defines these properties was examined. Another Ukrainian Standard (GOST) 3476-74 "Shot Blast-furnace and Electrochemical Phosphorus Slag for Concrete Production" was used for clarification of some slag properties which are significant for road building. Since the considered slag has been exposed at ambient air it has disintegrated to size fraction of 5 to 15 mm. This is favourable for the use of slag as a road base. The formula for calculating the so-called quality factor (K) from GOST 3476-74 was used for evaluating the properties of the slag grains including leachability. The fact that the content of MgO in the slag is lower than 10% was taken into account. This result met the requirements for the use of the basic slag as base material for roads. Since the price of such material is low and does not exceed 18-20 hryvnas (3.5-4 $ US) per ton, an important consideration in the cost effectiveness of selling it, is freight costs. The addition of freight costs can be a significant factor, depending on the location of the proposed road building relative to the plant in Konstantinovka.

Therefore, an economic analysis which fails to consider the effect of freight costs is not useful in arriving at a general conclusion about the expediency of slag usage. Each situation will have to be evaluated on a case by case basis, especially since the Ukrainian Government is changing the railway tariffs.

The more difficult issue concerns the problem of utilization of the acidic slag generated in the current technological process. The extra high SiO2 content forces the application of this waste in a silica–containing material. There are many very pure silica deposits in the USA, which are mined for foundry sand and glass production. Silica is cheap. There is very little acidic slag waste generated in the US. It is doubtful that anyone recovers silica from slag. Dumping charges for non-hazardous wastes are relatively cheap in most areas of the USA, and much the same situation exists in the Ukraine. The exception in the US is near big cities and the West Coast. There are no major steel mills or foundries left in these areas. So, acidic residues are usually just hauled to landfills.

Means of Saving Electric Power and Reducing Graphite Electrode Consumption

The work on improving the level of energy and graphite consumption was performed along two lines: firstly, the implementation of automatic coordination of the electrode operation, and secondly, the introduction of the electric current melting of oxidized fine particle Ni-containing wastes while minimising the electric arc occurrence.

ELECTRODE CONTROL

The VtorMet Company has introduced an automatic system of coordination of the graphite electrodes at the EAF. The main operating principles of this system are the switching on of the electric arcs at the start of melting and the controlling of open-circuit or short-circuit failures. Modulating regulation of the arc current was sufficiently reduced practice. Furnace operation within allowable limits of deviation from controlled parameters was achieved. The response time that eliminates short-circuit or open-circuit failures of the electric arc is 1.5 – 2.5 sec. So the number of times the high-voltage is switched off is reduced. The speed of movement of the electrodes is 2.5-3.0 m/min under the operating conditions. Since the length of an arc is minimal during the period of melting a solid charge, the speed of electrodes movement is minimised during the other periods of a melt. Therefore, the movement of the electrodes is unnecessary during the unsymmetrical changing or short-term fluctuation of the electric conditions that last only for a split second.

Every phase of the electrode operation is provided by an independent system of control. So all phases operate separately. The option of prompt conversion from automatic to manual regulation ensures the necessary indirect operations. The system for electrode movement includes a DC motor with separate excitation, a reduction gear, a mechanical transmission and a load-carrying structure that supports the electrode. These motors are characterized by a long starting period, a broad scope of operation, convenient speed control and the option of an easy-to-use dynamic slowdown, backspacing and setting.

Consequently, the produced and installed automatic control device provides continuous process control during the melt. It reduces melting time, energy and graphite electrode consumption due to the creation of optimum energy distribution between the arcs, and stabilization of temperature conditions and metal chemical composition.

ELECTROSLAG REMELTING

A bench-scale trial was carried out with the goal of minimising energy consumption by the application of the required electric controls with an electroslag remelting furnace with non-water-cooled crucible while melt refining Ni-bearing spent catalysts. The waste contained 94% Al_2O_3 as a carrier with 6% NiO as catalyst. The main goal of this melt was to determine the lining life. So different types of bricks were used as a lining material, namely magnesite, dinas and graphite. SiO_2 additions were introduced into the melt to lower the melting point of the slag. This was done with an electric current of 1000 A and up to 80 V voltage. This condition was controlled by the submerged depth of the electrode. The period of unfavourable electric arc occurrence was

172

minimised and limited by application of the "cold start" regime in the furnace. This regime requires increased energy and graphite consumption compared with the electric arc regime normally used at VtorMet. Graphite bricks were determined to have the best lining performance, showing that graphite is the best material for the lining. The master alloy produced contained: 70.6% Ni; 24% Al; 4.2% Si, and Fe and minor elements balance. The slag generated was nearly free of metal beads. This fact indicated the high metal extraction ratio – 98%.

Action plan for costly potential improvements

As the main result of the project, an action plan was worked out, listing costly potential improvements, which will lead to efficient and clean recycling production. This plan took into account the necessity of achieving the economical efficiency of the investments which have to be made in order to implement the necessary improvements.

THE PURCHASE AND INSTALLATION OF AN AIR POLLUTION CONTROL SYSTEM

Based on a review of the literature, discussions with control equipment vendors, and a review of control equipment installed at EAFs in the US, the project team concluded that a baghouse would be the most appropriate air pollution control equipment for the VtorMet plant. The total cost associated with acquiring a used air pollution control system was evaluated as being preferable for VtorMet to the purchase of a new system. It was determined that the borrowed capital necessary for the purchase would be returned in the first year, if 3,500 tons hazardous Ni-bearing wastes (stainless steel pickling, sludges, spent catalysts etc.) were recycled rather than landfilled. The cost of the 15% ferronickel produced would cover the expenditure on production here. VtorMet would also get $175,000 US, which waste generators would have to pay to landfill this amount of Ni-bearing wastes. The calculation does not include savings of reusing the collected dust by the installed air pollution dust collection system (waste returning). These additional savings should outweigh the disadvantage of a possible fall in the LME Ni-price, such as has been seen recently.

SLAG RECYCLING

Major attention will be paid to the recovery of the metal beads, which average up to 12 % of the total slag volume. This will include both currently generated slags and previously landfilled slags on the VtorMet site. The experience gained at the Russian plant Electrostal when faced with the same problem, is evidence of the cost effectiveness of this activity [1]. Rough estimations were made to consider the option of recovering beads on a toll basis for the Azovstal Works which are located close to VtorMet. Selling beads as stainless steel scrap as well as selling the balance of the basic slag to road builders was taken into account. The cost effectiveness of this business is approximately $ 2.0 US per ton of slag calculated on the basis of the current stainless steel scrap price in the Ukraine and after deduction of recycling and freight costs. The average metal extraction rate obtained at Electrostal – 30 kg per 1000 kg of slag - was used as the basis for the calculations.

When recycling acid slag, the utilization of the non-metallic portion is not anticipated. It would be landfilled on-site at the VtorMet plant. Nevertheless, the cost effectiveness should be close to the previous calculation since the price for the non-metallic portion is very low and no off-site location would be required for the landfill operation. The company will have unfilled space for this purpose on its own site after the removal of the landfilled basic slag.

Since the rate of slag recycling will exceed the rate of slag generation, the proposed activity will lead to a net decrease in VtorMet's landfill area. Cost effectiveness of slag recycling and use of the cleared VtorMet grounds for other purposes will clearly be of economic benefit to the plant.

INTRODUCTION OF ELECTROSLAG MELTING WHEN REFINING NI-CONTAINING FINE WASTES

Electroslag melting conditions when applied in melting fine wastes containing nickel are characterized by reduced energy consumption in comparison to standard electric arc melting and refining [2]. The anticipated energy saving is 270 kWh per ton of recycled waste. This figure was obtained from a comparison of the minimal achievable electric power consumption (1130 kWh) and the average for the Ukraine (1400 kWh). Expenditure on the introduction of this melting condition arises principally from the increase in graphite materials consumption of 1.7 kg per ton of waste. It was calculated that the cost saving for the introduction of the electroslag technology is $ 5.36 US per ton of recycled waste.

Conclusions

It was demonstrated that good air pollution control equipment can be found in the used market that will allow one to significantly improve environmental situations without the expense of new equipment. In addition, the project demonstrated pollution prevention, in this case by dismantling the batteries and removing the plastics before charging to the furnace. This project demonstrates to all similar industrial operations in the Ukraine that, if their process is carefully analysed with respect to pollution prevention and pollution control, relatively simple changes can be made which will yield significant improvements to the atmosphere around these facilities and ultimately improve the health and lives of ordinary citizens in the Ukraine.

This project did demonstrate that the necessary pollution controls could be implemented without significant impact on the financial status of the company. The abilities of pollution measurement experts, and the abilities of other experts in melting furnace design and automation in the Ukraine, prove more than adequate in solving the problems.

The visit to melting facilities in the USA provided valuable insight into ways in which meaningful improvements can be made at reasonable cost.

Considerable electric power savings should be achieved by applying electroslag melting conditions in a furnace when recycling fine Ni-containing wastes.

Graphite electrode conservation has been achieved by introducing an automatic system of electrode coordination in the VtorMet furnace.

The most profitable operation in recycling basic slags from stainless steel productions is recovery of metal beads, which may be up to 12 % of the total slag volume. The remaining non-metallic portion meets Ukrainian standards as a road-building material.

174

Acknowledgements

The work was conducted with the financial support of USAID. Special thanks are also extended to EcoLinks officers Mr. Yesirkenov and Ms. Schetinina for constant assistance in all the project's activities.

References

Galkin M.P, Larionov V.S., Stepanov A.V. and Nikitin G.S. (1998) Recycling landfilled slags at metallurgical works, Tekhnologicheskoe Oborudovanie I Materialy, 4,1998, pp 36-37.

Sunnen J. (2001) Industrial recycling by the electroslag process, Proceedings of the International Symposium on Electroslag Remelting Technologies and Equipment, pp 165-167. Kyiv: Elmet-Roll, Kyiv.

CHAPTER 18

Formation and Valorisation of Blast Furnace Mud from Usage and Treatment of the Industrial Water in the Iron and Steel Corporation Sartid A.D. – Smederevo

Dr. Ljubomir Sekulic, Dr. Predrag Lalic

SARTID Institute for Metallurgy d.o.o.
Goranska 12
11300 Smederevo
Serbia and Herzegovenia

Summary

This chapter deals with the supply and treatment of the industrial water in SARTID's production units, and analyses the quantities and chemical characteristics of waste water. It also presents the results of a survey of the volume and treatment of mud and sludge.

Water Supply and Use in Sartid A.D.

Taking into account technological processes and operations, and bearing in mind the required quality of water, pressure, physical and chemical properties of water before and after use, SARTID has adopted a process whereby the water supply system is divided into the flow through type A systems and type B re-circulation systems.

Special attention has been paid to the following: all required water used in the technological process of SARTID a.d. can be taken from the River Danube.

Water from certain technological processes (cooling of compressors and condensers in the utility plant) can be reused without any cooling or purifying, as it is warmed up by just 7-8°C. This reduces the costs of operation and investment.

Plants using the industrial water only for cooling of machinery and plants, and where it is technically not feasible to introduce re-circulation systems, discharge water through collector II into the River Ralja. This water is not polluted, but only warmed up by several degrees Centigrade. About 5,500 m³/h are discharged both when only the blast furnace VP-2 is in operation and also when the blast furnaces VP-1 and VP-2 are in operation at the same time.

In plants using the industrial water for cooling, and where technically feasible, re-circulation systems have been built in, and used water is returned into the process via the cooling towers (re-circulation system B). When only the blast furnace VP-2 is in operation, about 21,000 m³/h of water flow through the re-circulation system B, and when both blast furnaces are in use, over 33,000 m³/h of water flow through this system.

For plants where water is mechanically or chemically polluted because of the technological operations it is used for, the type A re-circulation system has been adopted. Discharging this water

W. Leal Filho and I. Butorina (eds.),
Approaches to Handling Environmental Problems in the Mining and Metallurgical Regions, 175–179.
© 2003 *Kluwer Academic Publishers. Printed in the Netherlands.*

through collectors I and II would not be permissible due to the physical and chemical characteristics of water and the hydrological characteristics of the recipient. For these reasons, plants for the mechanical processing of this water to a degree which would make reuse possible, have been provided. Water is purified in Dor thickeners of 30 m in diameter and peripheral grabs (Agglomeration – 2 thickeners, blast furnaces – 3 thickeners, B.O.F. Steel Plant – 4 thickeners), five horizontal settlers 10 x 50 m in size and 10 sand filters in the hot rolling mill. The cold rolling mill is provided with a plant for chemical treatment of the water used in the chemical treatment of metal surfaces (pickling, electrolytic degreasing). This has a capacity of 106 m^3/h. In plants where, in addition to physical and chemical pollution, the temperature of the water is raised, cooling towers are provided to cool the water down to the required temperature that would allow its reuse.

Taking the above facts into consideration, SARTID a.d. has adopted a water supply system divided into the following sections according to the method of supplying the consumers:

- A fresh water supply system – the through flow system;
- A re-circulation water system – the return water using system.

The consumption of water per production plant is shown in Tables 1 and 2, and the volume of water discharged into the River Ralja through collectors I and II is given in Table 3.

Table 1 – SARTID a.d. water consumption (two BFs in operation (1,800,000 t/year of final products)

PLANT	Re-circulation systems (m^3/h)		Flow-through	Total
	A	B	m^3/h	m^3/h
Tehnogas	-	4,019	-	4,019
Cold rolling mill	-	5,025	106	5,131
Hot rolling mill	10,422	2,000	-	12,422
Cut-up shear in hot mill	-	-	150	150
Compressor station	-	-	600	600
B.O.F. steel plant	8,500	5,000	2,320	15,820
Agglomeration 1 and 2	1,371	-	1,600	2,971
Blast furnace 1 and 2	2,012	5,500	-	7.512
Utility plant	-	12,000	-	12,000
Chemical water treatment plant	-	-	350	350
Granulation of BF slug	540	-	-	540
Casting machine	100	-	-	100
TOTAL:	22,945	33,544	5,126	61,615

Table 2 – SARTID a.d. water consumption (with BF-2 in operation) (1,000,000 t/year)

PLANT	Re-circulation systems (m³/h)		Through-flow m³/h	Total m³/h
	A	B		
Tehnogas	-	-	2,400	2,400
Cold rolling mill	-	5,025	106	5,131
Hot rolling mill	10,422	2,000	-	12,422
Cut-up shears in hot mill	-	-	150	150
Compressor station	-	-	600	600
B.O.F. steel plant	8,500	5,000	2,320	15,820
Agglomeration 1 and 2	750	-	800	1,550
Blast furnace 1 and 2	1,006	3,100	-	4,106
Utility plant	-	6,000	-	6,000
Chemical water treatment plant	-	-	200	200
Granulation of BF slug	240	-	-	240
Casting machine	100	-	-	100
TOTAL:	21,018	21,125	6,575	48,719

Table 3– Water discharge through collectors I and II into the Ralja River

PLANT	Collector I m³/h	Collector II m³/h	TOTAL m³/h
Tehnogas	-	2.400 (-)*	2.400 (-)
Cold rolling mill	-	106 (106)	106 (106)
Cut-up shears in hot mill	-	150 (150)	150 (150)
Compressor station	-	600 (600)	600 (600)
B.O.F. steel plant	-	2,320 (2,320)	2,320 (2,320)
Agglomeration	-	800 (1,600)	800 (1,600)
Utility plant	110 (110)	-	110 (110)
Chemical water treatment plant	14 (14)	-	14 (14)
Excess water from hydro-column	940 (940)	-	940 (940)
TOTAL:	1,064 (1,064)	6,376 (4,776)	7,440 (5,840)
*The volume of water during parallel operation of blast furnaces (BF-1 and BF-2)			

As can be seen from Table 3, with only Blast Furnace 2 in operation, about 48,700 m³/h of industrial water is consumed, or about 380 m³ per ton of produced steel.

Of this volume, about 6,500 m³/h of water is discharged, or about 13% of the total amount of water used, and in this respect SARTID a.d. does not lag behind other world steel plants throughout the world. About 42,000 m³/h of water is in re-circulation, about 21,000 m³/h being in the type A re-circulation system (water polluted mechanically or chemically), and about 21,000 m³/h in the type B re-circulation system (this water is not polluted and is returned into the process via the cooling towers). During parallel operation of BF-1 and BF-2, the total volume of the industrial water will exceed 61,600 m³/h, with the volume of water in type B re-circulation system considerably increased, whereas water in type A re-circulation will remain practically unchanged (Table 2).

All re-circulation systems are fed with industrial make-up water from the channel via the pumping station Lipe (table 4). The volume of make-up water varies depending on the system, but in no case does it exceed 10% of the total volume of water in re-circulation.

Table 4 – Distribution of 10,800 m³/h of water from the pumping station Lipe

PLANT	Re-circulation System Make-up		Flow-through m³/h	Total m³/h
	A	B		
Tehnogas		- (400)	2,400 (-)	2,400 (400)
Cold rolling mill		454 (454)	106 (106)	560 (560)
Hot rolling mill	850 (850)	-	-	850 (850)
Cut-up shears in Hot Mill	-	-	150 (150)	150 (150)
Compressor station	-	-	600 (600)	600 (600)
B.O.F. Steel Plant	855 (855)	-	2,320 (2.320)	3,175 (3.175)
Agglomeration 1 and 2	75 (150)	-	800 (1,600)	875 (1,750)
Blast Furnace 1 and 2	100 (200)	300 (500)	-	400 (700)
Utility plant	-	-600 (1,.200)	-	600 (1,200)
Chemical water treatment	-	-	200 (350)	200 (350)
Granulation of B.F. slug	50 (100)	-	-	50 (100)
Casting machine	-	-	940 (940)	940 (940)
TOTAL:				10,800 (10,775)

Sludge and Scale Treatment

From the water treatment system, over 100,000 tons of sludge and scale, or about 100 – 120 kg/t of steel, are separated annually. The separated material is returned into the process of sintering as it contains 45 - 65% of iron. An exception to this is the blast furnace sludge, which, in addition to 45 - 50% of Fe, contains also about 10 – 15% of Zn, so that it cannot be returned into the process. This sludge is used for the production of fertilizers with microelements (Zn, Fe, Mo, Ti, etc.) at IHP Prahovo.

Physical and Chemical Characteristics of Waste Water in Sartid

From the physical and chemical characteristics presented in (Table 5), it can be seen that, after the wastewater is discharged into the river, water in the river becomes more turbid and contains more suspended and sediment matter, grease, oil and ammonia, and has a higher concentration of iron, manganese, copper and zinc ions.

Table 5 – Physical and chemical characteristics of the waste water

PARAMETER	Upstream 100 m	Collector I	Between collectors	Collector II	Downstream 100 m
Visible waste matter	-	-	-	-	-
Water temperature, °C	17	21	17.5	23.0	18.5
Turbidity in degrees	30	40	30	300	150
Colour in degrees	15	20	20	20	20
pH value	8.33	8.49	8.45	8.31	8.35
Nitrates, N_2O_4, mg/l	3.0	3.5	3.0	7.00	5.00
Nitrites, N_2O_3, mg/l	0.05	0.05	0.03	0.03	0.04
Ammonia, NH_3, mg/l	1.5	2.5	1.5	3.0	2.0
Chlorides, Cl, mg/l	34	66	40	82	55
Consumption of $KMnO_4$, mg/l	23.38	21.52	17.76	17.36	23.86
HPK, mg/l	5.6	4.20	6.50	4.20	5.38
BPK, mg/l	2.74	-	-	-	3.05
Alkalinity, ml/l o.1N HCl	73	60	78	34	44
Iron, Fe, mg/l	0.3	0.8	0.3	5.2	1.8
Manganese, Mn, mg/l	0.10	0.70	0.3	0.5	0.15
Sulphates, SO_4, mg/l	65.92	56.32	62.72	30.8	34.56
Phosphates, PO_4, mg/l	0.70	0.00	0.50	0.50	0.75
Oxygen, O_2, mg/l	10.42	-	9.81	-	8.36
Suspended matter, mg/l	10	11	16	234	16
Sediment matter, mg/l	0.2	0.1	0.2	0.7	0.4
Detergents, mg/l	0.00	0.00	0.00	0.00	0.00
Grease, oil, mg/l	0.00	0.001	0.001	0.07	0.06
Phenols, mg/l	0.00	0.00	0.00	0.00	0.00
Lead, Pb, mg/l	0.00	0.00	0.00	0.00	0.00
Zinc, Zn, mg/l	0.00	0.021	0.016	0.055	0.036
Cadmium, Cd, mg/l	0.00	0.00	0.00	0.00	0.00
Copper, Cu, mg/l	0.00	0.021	0.013	0.030	0.025
Chromium, Cr^{+6}, mg/l	0.00	0.00	0.00	0.00	0.00
Calcium, CaO, mg/l	162.57	-	176.58	-	112.11
Magnesium, MgO, mg/l	188,63	-	90.65	-	32.18

The pH value of the water remains unchanged before and after discharge of wastewater from SARTID a.d., and is within the permissible range. No presence of phenol, cadmium, lead or six-valence chromium has been found. The HPK and BPK values remained unchanged before and after discharge of wastewater from SARTID a.d.

CHAPTER 19
The usage of pneumatic conveying systems in handling environmental problems in the metallurgical and mining industry

Dr. Jens Lahr, Werner Kaulbars

E.S.C.H. Engineering Service Center und Handel GmbH
Maxhuettenstrasse 19
D-07333 Unterwellenborn
Germany

Summary

Pneumatic conveying or metering technologies are important for the ecological progress of metallurgical processes in several respects:

- Decreasing of pollution
- Decreasing of energy consumption by electric arc furnaces (EAF) as a result of slag foaming
- Decreasing of dust pollution in casting machines by means of granulated flux powder metered automatically by a pneumatic conveying system directly onto a mould
- Substitution of metallurgical coke by injecting coal powder into the tuyeres in blast furnaces (BF)
- Use of waste as raw material or energy source
- Substitution of metallurgical coke by injection of waste materials, such as processed rubbish or dried settling sludge into the tuyeres of BFs and cupolas
- Injection of dusts into steel baths of converters (BOF) and EAF
- Waste heat cleaning with the help of injection of high diluted additives such as coke and/or lime powders for the adsorption of dioxins and furans

This chapter reviews the use uf pneumatic conveying systems and concludes that the reliability of the relevant technical applications has become high enough (more than 99 % of operation time) for widespread use.

Introduction

E.S.C.H. was established in 1992. The aim of the company is to make use of existing experience in application of pneumatic conveying in the iron and steel industry, to evolve new technologies and to employ them in this and other industries. We determine the most convenient process technology for the task in question. We utilize anything from a well-equipped laboratory to homemade software as the problem in hand requires. The custom-built systems are installed at the customer's plant, put into operation and finally supervised by E.S.C.H.-service. Over the last 10 years several metering, injection and conveying systems have been erected and put into operation at blast furnaces and sintering plants, in steel shops, at cupolas, at aluminium melting furnaces and in other industrial plants. The experience gained shows that the majority of the tasks to be solved are directly or indirectly connected with improving the environmental situation of the basic technologies,

181

W. Leal Filho and I. Butorina (eds.),
Approaches to Handling Environmental Problems in the Mining and Metallurgical Regions, 181–187.
© 2003 *Kluwer Academic Publishers. Printed in the Netherlands.*

especially in the steel industry. This paper gives some examples of successful use of pneumatic conveying for the improvement of metallurgical technologies from the environmental point of view.

Reducing pollution

The following technical solutions help to reduce pollution either directly or indirectly by reducing the consumption of energy or raw material.

Slag foaming

Consumption of electric energy in electric arc furnaces (EAF) depends on the heat losses of the steel bath. This heat loss occurs principally as a result of radiation heat transfer from the surface of the steel bath to the water-cooled roof. Injection of coal powder into the slag has become an efficient technology to reduce such heat losses by foaming the slag. Hence, the foamed slag works as heat insulation.

The technical task of the coal injection equipment is to make the following possible:

- Injection through several metering lines, whereby each line can be operated independently of the others;
- Injection with a wide range of mass flow rates;
- High availability

For the technology of multiple metering developed for this purpose, pneumatic methods only are used. The pneumatic control of coal mass-flow without any mechanical metering device guarantees the high availability of nearly 100 % needed for efficient operating of EAF. Fig. 1 shows the lower part of the feeder for slag foaming at an 80 t-DC-EAF.

Fig 1. Lower part of the feeder for foaming coal with 2 outlets and feeding lines

The 2 feeding lines (1 for the manipulator and 1 for the side wall injection lance) can be independently operated. Each of them is able to inject 10-120 kg/h. The internal diameter of the feeding pipelines is 40 mm.

Flux powder metering

Work places at continuous casting machines are polluted both by fume generated by molten steel and slag and by dust generated by handling flux powder. SiO_2-dust is the most dangerous component which can be contained in flux powder. That is one of the reasons why granulated flux powder has been used increasingly over the last few years. Furthermore, the granulated flux powder makes it possible to automate metering of the flux powder to the surface of the steel bath.

A number of systems for metering granulated flux powder were developed. The common disadvantage of these is that sufficient space is needed between casting platform and tundish and/or in front of the casting platform. This condition is frequently not fulfilled at slab casters. This is the reason why the system shown in Fig. 2 was developed.

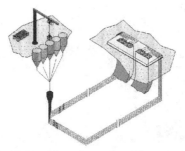

Fig 2. Scheme of automatic metering of granulated flux powder directly onto the mould

Granulated flux powder is metered through a number of conveyor lines directly onto the mould [1]. Pneumatic stopper conveying technology is used for this purpose. Each line can be independently operated. Metering is realized by controlling the duration of the feeding periods. The pollution of dust generated by granulated flux powder conveyed directly onto the mould is lower than by manual handling of granulated flux powder.

Fig. 3 shows 1 of altogether 6 outlets of the conveying lines to slab caster with adjustable mould. The number of operated conveying lines (2, 4 or 6) depends on the adjusted width of the mould.

Fig 3. Mould with outlets of granulated flux powder

184

Of course, installations for metering of granulated flux powder directly onto the mould can also be used at other continuous casting machines such as bloom casters, beam blank casters and billet casters.

Injection of coal powder into BF

Coke production is one of the most important sources of pollution as far as the production of hot metal in blast furnaces (BF) is concerned. So, the substitution of metallurgical coke by injection of coal powder into the tuyeres reduces not only the costs of the hot metal but also combats the summary pollution ensuing in hot metal production. KOSTE-Technology, developed by Maxhuette Unterwellenborn and put into operation for the first time in 1982 [2], is one of the technologies used today. It is based on the dense phase conveying technology. The metering pipelines (Fig. 4) are installed separately, for each tuyere, from the metering feeder to the tuyeres of the BF, and are separately and independently operated.

Fig 4. Metering pipelines from KOSTE-plant to BF in Ling Yuan, P.R. of China

Pneumatic methods only are used to control the mass flow rate. This ensures a high rate of availability, amounting to nearly 100% of the operation time of the BF.

In addition to coal powder, other solid fuels such as dried settling sludge and processed rubbish were considered for use as additional fuel in BFs. From the point of view of pneumatic conveying, these materials are basically different from coal powder in that they are coarse-grained. This means that such materials cannot be used in dense phase conveying. Hence it was necessary to develop technology suitable for coarse-grained materials, working on the basis of dilute phase conveying and performing in the same way as the KOSTE-technology (independent operation of the metering lines, no mechanical metering devices) [3]. This development was supported by "Deutsche Bundesstiftung Umwelt". A metering plant (Fig. 5) has been erected, which contains equipment both for powder injection and for the injection of coarse grained materials into 13 of the 15 tuyeres at BF No. 3 of EKO, Eisenhuettenstadt.

It takes only 8 hours to adjust from one feeding system to the other.

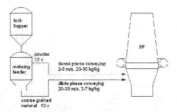

Fig 5. Combined KOSTE-plant showing installation for the alternative injection of coarse grain materials

Use of waste as raw material or energy source

One of the most efficient means of decreasing the amount of waste produced by industry and households is the utilization of such wastes as raw material and/or as energy sources. This reduces not only the cost incurred in disposing of waste at incinerating plants or rubbish dumps but also reduces costs for raw materials and energy.

Injection of waste material into blast furnaces (BF)

A percentage of metallurgical coke used in BFs and cupolas can be substituted by natural fuels such as coal powder, natural gas or heavy oil, but it is also possible to use artificial fuels such as processed rubbish (Fig. 6) or dried settling sludge [4,5].

Fig 6. Processed rubbish, grain size <10 mm

Processed rubbish contains about 90 % plastics. It has a grain size of up to 10 mm. This makes metering and distributing by means of pneumatic conveying much more complicated than the metering of pulverized materials such as coal powder. The KOSTE-technology developed and modified for use with coarse-grained materials is suitable for such material. Fig. 7 and fig. 8 show the plant used for injection of processed plastic waste in BF No. 1 at EKO Stahl, Eisenhuettenstadt, which has been in operation since June 2001.

Fig 7. Lower part of the feeder for injection of coarse grain materials into 13 BF-tuyeres

Fig 8. 13 metering pipelines near the feeder for injection of coarse grain materials

In the first year of operation the following parameters were reached:
- Availability: 98 % of the operation time of the BF
- Average specific mass flow of plastic waste: 67 kg/ t h.m.
- Coke substitution ratio: 0.92 kg coke / kg plastic waste.

Injection of dusts into steel baths of BOF and EAF

Injection of dust from the waste heat cleaning of converters and EAFs into steel baths is used for the concentration of ZnO in the dust. At the same time, the iron combinations present in the dust are utilized and the other materials are slagged. This technology makes sense if dust with a high concentration of ZnO can be efficiently sold to the zinc industry.

Waste heat cleaning

Treatment of waste heat using injection of high-diluted surface active additives such as pulverized lignite hearth type furnace coke and/or stone or slaked lime powders is increasingly applied in the cleaning of toxic substances such as dioxins and furans. Additives have to be metered to the waste heat stream with a relatively low mass flow rate. Naturally, even distribution is important for the effect of the injection. For this purpose a number of injection lances are placed in the off-gas conduit (Fig. 9, 10).

The additives are distributed using a static distributor (Fig. 10). Such a distributor does not contain any mechanical metering devices, which is the main reason for the high availability attained.

187

Fig 9. Waste heat conduit of EAF with injection lances for additives

*Fig 10.*Distributor and injection lances on the roof of the waste heat conduit of sinter plant

References

Lahr J., Kaulbars, P., Hirsch, R., Kempe, W., Tabor, R., Buchwalder, J. and Hunger, J. (2000) Entwicklung und Erprobung eines Verfahrens zur Verwertung und thermischen Nutzung von Rest- und Abfallstoffen aus der Abgas- und Abwasserreinigung durch Einblasen in metallurgische Schmelzreaktoren: Abschlussbericht zur Phase 3 des Vorhabens, Deutsche Bundesstiftung Umwelt, E.S.C.H. GmbH, Unterwellenborn.

Letzel, D. (1996) Kinetik der Verbrennungsvorgaenge von Kohlenstaeuben und Reststoffen beim Einblasen ueber die Windformen des Hochofen. Dr.-Ing. thesis, Technische Universitaet Bergakademie Freiberg.

Hunger, J. (1997) Einsatz fester Brenn- und Reststoffe ueber die Windformen des Hochofens. Dr.-Ing. thesis, Technische Universitaet Bergakademie Freiberg, 1997.

Ortloff J., Kaulbars W., and Lahr J., Verfahren zur Zufuehrung von Giesspulvergranulat zu einer Stranggusskokille, Patentschrift DE 199 56 059, Int. Cl.: B22 D 11/07

Scheidig K., Guether, R., Engel, G., Vogel, M., Kretschmer, H., Schingnitz, M. and Goehler, P. (1985) Betriebserfahrungen mit einer neuen Generation von Anlagen zum Kohlenstaubeinblasen in den Hochofen, in Stahl und Eisen, Vol. 105 (1985), pp. 1437-1441.

CHAPTER 20

Use of the melting-gasification process for processing metallurgical sludges

Renate Lahr[1], Klaus Scheidig[2], Hans Ulrich Feustel[3], Joachim Mallon[3], Michael Schaaf[3]

[1] E.S.C.H GmbH Unterwellenborn, Maxhuettenstrasse 19, , D-07333 Unterwellenborn, Germany
[2] VTI Verfahrenstechnisches Institut für Energie und Umwelt, Box 2005, 07318 Saalfeld, Germany
[3] Ingitec Ingenieurbüro für Gießereitechnik GmbH, Box 202, 04167 Leipzig, Germany

Summary

Ingitec Leipzig have been successful in developing, in co-operation with partner firms, the IMeGa-S – process (Ingitec-Melting-Gasification with Special Solutions). This process represents a fixed-bed melting-gasification in parallel flow, injecting industrial quality pure oxygen. The process is applied to produce synthesis gas for power and heat generation from waste materials and/or biomasses. The synthesis gas analysis ranges from 40-55% CO and 15-25% H_2, depending on the material gasified and the process control mode.

Depending on the input material the IMeGa-S - reactor produces some slag and hot metal, which must be tapped from time to time or, if the capacity is high enough, even continuously.

The technology developed guarantees the complete destruction of any organic pollutants possibly included in the material charged under reducing conditions and at high temperatures up to 2000 °C. The range of charges covers plastic waste, municipal waste, sewage sludge, shredder fractions, old tyres, waste wood, natural biomasses, etc.

Operating in an experimental plant near Leipzig (Germany), the IMeGa-S - process has been developed using a small-scale reactor (hearth diameter: 800 mm; throughput: max. 1.5 t/h). The first commercial 2sv-plant is under construction, but is very delayed. The plant is being designed with a capacity of 50,000 t/a of old timber; electrical power: 7.5 MW; thermal power: 9.4 MW.

Description of the oxygen-melting-gasification method

The IMeGa-S - process is based on the metallurgical process technology of the oxygen cupola furnace [1] - [5]. The reactor developed on this basis is characterised by two tuyere levels, with the gas exhaustion chamber positioned between them. The hearth of the reactor is lined with refractory material up to the height of the bottom tuyere level (level 2). It is provided with a double-wall, water-cooled steel jacket (Figure 1).

The particular feature of this new process is that the reaction gases are passed in parallel flow with the input materials in the shaft of the reactor. The advantages of material and heat transport in counter flow are deliberately ignored here, in order to ensure destruction of organic contaminants under reducing conditions and at high temperature. The gaseous reaction products only move in

189

W. Leal Filho and I. Butorina (eds.),
Approaches to Handling Environmental Problems in the Mining and Metallurgical Regions, 189–195.
© 2003 *Kluwer Academic Publishers. Printed in the Netherlands.*

190

counter flow and/or cross flow towards the movement of the packed bed in a relatively small volume within the hearth of the reactor.

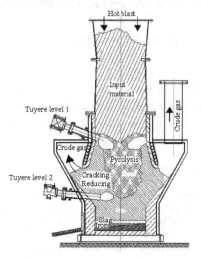

Fig. 1. Schematic representation of the IMeGa-S - reactor

The input materials are fed via a hopper and a gate into the gasification shaft. The cold material is dried and pre-heated by hot blast introduced from the top of the reactor. A high-temperature zone is formed at the upper tuyere level (level 1) where industrial quality pure oxygen is injected. Within this zone the organic components contained in the input materials are partially burnt by means of the oxygen. Depending on the material type, extent of pre-treatment, drying and pre-heating, the oxygen used and other influential factors, different combustion levels are generated with combustion temperatures of up to 2000°C. The hot gaseous reaction products are exhausted downwards in parallel flow, together with the descending input materials. The mineral and metallic components of the input materials melt. They drop or flow downwards and are collected in the hearth. The organic components of the input materials which are not burnt completely in the tuyere level 1, to one extent are gasified, in part being subjected to pyrolytic decomposition or cracking.

The pyrolysis coke produced reacts with oxygen in the lower tuyere level (level 2). The heat generated here is used to compensate the heat and temperature losses due to the endothermic reactions between the 1st and 2nd tuyere level (gasification, pyrolysis and cracking of organic components as well as reduction of the combustion products from level 1). Thus, it is ensured that slag and metal can remain fluid in the hearth and be tapped or flow off continuously in large quantities.

The crude gas exhausted from the zone between the two tuyere levels contains virtually no contaminated organic materials because neither organic compounds nor the gaseous pyrolysis products, produced from the latter, are stable at high temperatures or under reducing conditions. CH_4 contents of < 1 % in the gas produced are a significant indicator of this. The Boudouard equilibrium to high CO contents is provided for the same reason.

The crude gas is exhausted from the 2sv-reactor within the temperature range of 750° – 850 °C. It is cleaned by conventional processes. The first stage is a dry gas cleaning system consisting of a spray absorber and bag filter system. It is followed by a wet gas cleaning system consisting of quench cooling, aerosol separation and a 2-stage gas scrubber. The necessary negative pressure will be provided by an exhauster/compressor situated downstream, which finally presses the gas through an activated charcoal filter. The fact that oxygen is used, makes it possible to design all these apparatuses with relatively small dimensions.

Advantages of the oxygen-melting-gasification method

This new technology is characterised by low investment and operating costs and by the fact that it operates in an environmentally friendly way. It includes the following advantages for the process:

- Preparation of input materials is minimised. Coarse crushing to approx. 300 mm is sufficient. The proportion of < 10 mm fines should normally be below 10 %.
- Metallic or mineral components of the input materials need not be removed before charging (no sorting out by hand, no magnetic separation).
- Both organic materials and contaminated materials are destroyed by the process itself. No higher molecular hydrocarbons occur in the crude gas of the melter-gasifier. The De-novo synthesis of dioxins and furans is prevented by the process described.
- Well-known processes meeting highly developed standards can be used in gas cleaning. Gas cleaning requires relatively low investment because of the relatively small crude gas volumes per ton input and because contaminated materials are destroyed as a result of the process itself.
- Process-inherent destruction also applies to a number of inorganic and metallic materials. Metal oxides are reduced to a large extent, while highly melting heavy metals (e.g. Cu, Cr, Ni) are found entirely, or to a large extent, as alloy elements in iron, which in this way acts as a heavy metal sink.
- The other oxides remaining in the slag are eluate-safe bound in the glassy hardened slag matrix.
- An optimised process and product yield results from
- the generation of a fuel or synthesis gas which is best suited for application in a gas engine or as a substitute for natural gas as the energy carrier;
- the production of fluid slag which can be used in form of lumps or granules in the construction of streets and roads. Moreover, it is also suited for further processing by re-melting, casting, etc.
- the production of liquid iron which can then be used as scrap metal.
- A minimum of residues (normally up to 3 - 5 % by weight of the input materials) in the form of dust, sludge and/or salt, which have to be disposed of.
- To a large extent, the single-stage process and the automated operation ensure optimum employment of operators.
- A high specific capacity (approx. 2.5 t/h and m² of hearth area) and an extremely high energy conversion (up to 40 GJ/h and m² of hearth area) enable use of compact systems with low space requirements.

These advantages mean that investment costs are low, even for small capacities. For instance, the investment costs of a 50.000 t/a installation can be calculated to be lower than 250 Euro per ton and year. Thus, IMeGa-S - technology opens the way to profitable, decentralized small units for the thermal treatment of waste materials.

The pilot plant

The pilot plant was installed at Leipzig-Holzhausen to test further developments of the process, together with new input materials provided by customers. Hot commissioning was completed in November 2000. The pilot plant reactor is equipped with 3 tuyeres on each of the two tuyere levels. It has an effective hearth diameter of 0.8 m and, thus, an active hearth area of 0.5 m². At a throughput rate of up to 1.5 t/h, its rated capacity is 10.000 tonns per annum. Figure 2 shows a partial view of the melter-gasifier of the pilot plant.

We were able to obtain important information on the performance of the process from the pilot plant operation in 2001. This included the verification of characteristic figures, especially the high throughput rate of 2.5 t/h and m² of hearth area, which is approximately ten times higher than the corresponding data for grate incineration. In addition to the high throughput rate, there is a high power conversion of up to 40 GJ/h and m² of hearth area, depending on the material charged.

As with all shaft furnace operations, the use of coke is the prerequisite for normal melting-gasification. Coke provides the permeability of the packed bed, i.e. the gas flow through the material, especially at high temperatures, where it is the only solid constituent. Coke consumption ranges from 40 to 100 kg per ton of waste depending on the kind of waste material charged. Blast furnace coke can be used. However, best results were obtained using foundry coke 100 – 150 mm.

Fig. 2. View of the pilot plant reactor

Because of the acid constituents of the coke ash and depending on the chemical composition of the waste material, the addition of some limestone is necessary to achieve good flow characteristics of the slag.

The operation of the pilot plant is computer controlled. For this purpose a model was constructed and adapted to the results achieved during operation of the pilot plant. This provided us with first-rate experience in the evaluation of waste materials using the appropriate chemical analyses. Table 1 shows the operating conditions and firing capacity of the generated gas as determined for a tested mixture of input materials (melting gasification of old timber, ground wood rejects from the paper industry, bulky waste, and the high calorific fraction of municipal waste). The electrical and thermal capacities are related to the power generation from synthesis gas in a bio mass heating power station (BHKW) with gas engines.

Table 1. Model calculation of the IMeGa-S – process

Capacity : 6.58 t/h waste material	7 600 h/a	50 000 t/a
Old timber	304 kg/t	30.4 %
Ground wood rejects	220 kg/t	22.0 %
Bulky waste	290 kg/t	29.0 %
Municipal waste, high calorific fraction	180 kg/t	18.0 %
Metallic parts	6 kg/t	0.6 %
Auxiliary materials:		
Coke	40 kg/t	
Lime stone	40 kg/t	
Oxygen	274 Nm³/t	
Hot blast	300 Nm³/t	
Output materials: Crude gas	8 577 Nm³/h	
Slag	0,7 t/h	
Metal	0,2 t/h	
Synthesis gas parameters		
H_2	21.7 %	
CO	43.4 %	
CH_4	0.9 %	
CO_2	10.8 %	
N_2	18.1 %	
H_2O	5.1 %	
Calorific value	2.39 kWh/Nm3	
BHKW:		
Input	Output	
Gas volume consumed, dry 8 577 Nm³/h	Power, therm.	9.4 MW
Calorific value 2.39 kWh/ Nm³	Power, electr.	7.5 MW
Firing power 20.5 MW	Efficiency, therm. 45.9 %	
	Efficiency, electr. 36.6 %	

The IMeGa-S – process and applicability

The IMeGa-S - process is best suited to decentralised systems which are capable of utilising local waste potential in an economical manner. The advantages of the process specified above ensure that the specific investment costs of a turn-key system can be clearly kept below the amount of 500

Euro/t of annual capacity of input material. These costs should include all major supplies and services, from input material logistics to the tie-in of a BHKW, including the necessary engineering services, construction and auxiliary construction works, installation and commissioning.

The compact design and, hence, the fact that IMeGa-S – systems require only a small amount of space, ensure that it is easy to tie them in to existing infrastructures. This is particularly applicable in cases where the sale of the heat generated can be ensured. Other advantages which are independent of location, can be seen above all in the installation of relatively small systems at decentralised sites (use of local waste sources, low transport load, more readily accepted by local population.

An economical alternative should be considered if it is possible to ensure that the crude gas generated in the melter-gasifier is cleaned and then burnt as fuel gas in an existing boiler to substitute fossil primary energy carriers. The existing flue gas cleaning unit of the boiler system should then be used for final cleaning in order to ensure that exhaust cleanliness complies with regulations.

Further cost reductions should be possible by using appropriate local infrastructure. If we consider only the 2sv reactor with the tapping system for slag and iron, the gas cleaning system and the water treatment plant (without the reactor hall), investments could be reduced to less than 250 Euro/t annual capacity.

The Rothenburg bio mass power station project

The Rothenburg project was to be carried out by MFU GmbH Leipzig, a former partner of ingitec. The order covered the construction of a melting-gasification - system with an annual capacity of 50,000 t. It was intended that this bio mass power station be taken into operation by the end of 2001, but the project is a long way behind schedule because MFU became insolvent in summer 2001. Work has begun and should be completed in 12 months' time by another company.

The melter-gasifier in Rothenburg is designed for the thermal utilisation of old wood in accordance with the bio mass regulations. A total of 7.5 MW electrical power is to be generated by means of gas engines from the process gas produced. Power is to be fed into the mains according to the Renewable Energy Act (EEG). The waste heat generated when operating the gas engine will be supplied to the Rothenburg long-distance heating network.

Marketing

FMI Feld Maschinen- und Industriebau GmbH, Oer-Erkenschwick/Germany, are the new partners of Ingitec GmbH Leipzig. FMI constructs installations for the melting-gasification of waste materials using the IMeGa-S process with capacities ranging from10 000 t/a to 50 000 t/a.

Co-operation

For the melting-gasification of dust or sludge it is necessary to agglomerate the material. Special solutions in this field are provided by VTI Verfahrenstechnisches Institut Saalfeld, Germany, which has extensive experience in briquetting and pelletising of dry or wet fine grained material. VTI acts as a partner in co-operation with Ingitec and FMI.

References

Ruschitzka L. et al. (1994) Beitrag zur Theorie eines Kreislaufgas-Kupolofens. Giessereiforschung 46, 2/3, pp. 40-50

Ruschitzka L. et al. (1994) Das Kreislaufgas-Verfahren - eine alternative Variante des Kupolofenschmelzens. Giesserei 81, No. 10, pp. 297-303

Scheidig K. (1995) Roheisen aus dem Sauerstoff-Kupolofen als alternativer Einsatzstoff für den Elektrolichtbogenofen. Stahl und Eisen 115, No. 5, pp. 59-65, 177.

Scheidig K. (2002) In 3. Generation. UmweltMagazin Düsseldorf, No. 4/5, pp. 80-83.

Scheidig K. et al. (2002) Recycling and Waste Treatment by the Oxygen-Melting-Gasification Method. Paper presented at the Lulea Recycling Conference of MEFOS; TMS & Lulea University, Lulea/Se, 17-19 June 2002.

CHAPTER 21

Intensification of ventilation and minimisation of emissions in aluminium smelter potrooms

Dr. Szilárd Szabó

University of Miskolc, Hungary
Department of Fluid and Heat Engineering
H-3515 Miskolc-Egyetemváros
Hungary

Summary

The focus on environmental and labour hygiene issues has intensified in Eastern Europe in the past few years. On the one hand, minimisation and elimination of formation and propagation of both powdery and gas pollution in potrooms is a labour hygiene issue. On the other hand, the far-reaching unfavourable effects of pollution produced by industrial plants is an environmental issue. The knowledge of the propagation of solid and aeriform pollutants in air inside and outside of potrooms is of the utmost importance both for existing industrial plants and plants still to be designed. In the case of complicated and extremely variable climatic conditions, only well coordinated measurements combined with numerical simulation can result in accurate observation of natural ventilation and tracing of the emission of pollutants, changes in their concentration, and the location of their precipitation and condensation. As far as new construction or large-scale re-modification is concerned, this issue should be discussed while taking into consideration the employment of optimal ventilation of the building concerned.

This chapter reports on a project for investigating current conditions and recommending improvements to the ventilation of an industrial workplace. The Department of Fluid and Heat Engineering at the University of Miskolc, Hungary, is involved in the creation and development of the resources needed for the investigation. This paper comments on the equipment used and the results of the investigations carried out, and makes suggestions for the future options for development. The methods applied in the investigation will be demonstrated using the example of application to an aluminium smelter potrooms.

Introduction: methods and means of measurements

Tested buildings are generally both naturally and artificially ventilated. This means that places within the building where pollutants are concentrated have their own localised removal systems. The air full of drawn dust and poisonous gases is transported into a scrubber, where harmful substances are separated out. It is impossible to collect all dangerous substances from a huge potroom housing complicated technological equipment, although efforts should be made to do so.

The pollutants in the air of the potroom are drawn off by an ordinary ventilation system. In the case of a large-size potroom, however, these systems are only aimed at dilution of the pollution and its transportation out of the building and not at its separation. The pollutants transported out of the building are released to the environment and cause pollution. The main objective is to dilute the

197

W. Leal Filho and I. Butorina (eds.),
Approaches to Handling Environmental Problems in the Mining and Metallurgical Regions, 197–209.
© 2003 *Kluwer Academic Publishers. Printed in the Netherlands.*

emitted pollutants to the desired extent and evacuate them (through a stack) so that the measured pollution will not exceed permitted health limits.

Taking into consideration the above mentioned facts, it can be stated that the key factor in this issue is the minimisation of the emitted pollutants and an increase in the ability to collect them.

Several different aspects, all involving measurements of the air pollutants inside and outside the potroom should be considered. There are three main methods of measurements:

A. Local measurements on-site
B. Modelling in the laboratory
C. Computer simulation

Methods B and C can be combined or used separately. On-site measurements can not be neglected. However, the combination of all three methods gives the most reliable results.

Observation of the flow of the emitted pollutants in the potroom revealed that direct sucking and ventilation of the potroom should be investigated separately. These systems are independent and different, and thus must be discussed in different sections, which will deal with the various measurements and the above-mentioned methods (A, B, C).

Exhaust systems

In the case of an aluminium smelter potroom, harmful gases in the pot are removed by the method of connecting the pots to a burner. This burner adds air to the emitted gases on the ejector principle, burning the received mixture. The burner is connected to the exhaust or gas and dust removal system. The point is to exhaust the hot flue consisting not only of gas, but also of a huge amount of dust and tar. Figure 1 is a schematic depiction of the exhaust system. Only one of the original 176 pots is drawn, and it should be mentioned that the system is also connected with other pots. Most of the dust content in the flue is removed by a multi-cyclone. Next, harmful gases are separated by gas scrubber pipelines. At the beginning some alum is added to the gas. The dust binds the harmful components of gases. The dust gas enters a sack cyclone which separates out the fine dust. The gas cleaned of the harmful gases and dust, flows into the fan maintaining the flow. The gas leaving the fans enters the stack through sound absorbers. The loss of heat and the added alum cool the gas, which is evacuated through the stack.

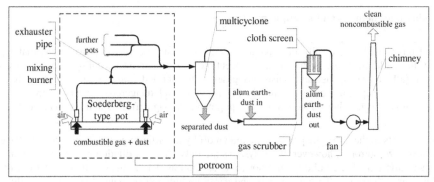

Fig.1. Block scheme of the exhaust system

Designing and operating such a complicated system requires special professional knowledge. The following sections deal with the processes involved.

Development of a burner

Our aim was to examine the burner heads connected to the pots in the laboratory, develop more efficient heads and analyse the factors which would ensure continuous burning, as the heads tended to cease burning, hence releasing some gas components into the exhaust system and later into the environment. There was another problem, namely the fact that the proportion of air-gas mixture was badly chosen. Thus, too much air entered the system and reduced the efficiency of the gas exhaust (see Szabó et al., Analysis and redesign...). A sample device was built to measure the flow in the burner and develop various burner types. Figure 2 shows this device (1- pressure fan, 2- valve, 3- orifice, 4- valve, 5-gas burner manifold, 6- gas burner pipe, 7- gas burner head, 8- flexible pipe, 9- suction fan, 10- flow nozzle, 11- valve). On-site measurements were carried out before modelling in the laboratory was done. As a result of our investigations, we found some basic relationship between the geometry of the burner and its fluid mechanics parameters. On the basis of our findings we made proposals regarding the construction of a burner ensuring stable burning and the correct proportion of air in the air-gas mixture. Manufacturing and testing of the burners is in progress.

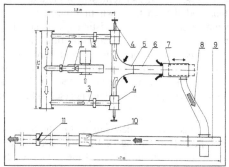

Fig. 2. Experimental device for measuring the flow in the burner and for developing various burner types

Redesign of the exhaust system

There are 176 pots, each with a burner at either end. They constitute the beginning of the exhaust system. Thus, there are 352 exhaust pipes altogether. There are four rows of pots (marked A, B C, D), and within these there are four sectors with 10-13 pots each. Figure 3 gives details of a section of the system shown in Figure 1. The squares in Figure 3 represent sectors. The rows and the number of pots are indicated in the squares.

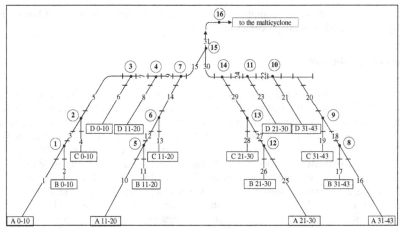

Fig. 3. Section of the exhaust system before the multicyclone shown in Figure 1

Looking at Figure 3 it can be seen that the system, which includes elements such as a system of cyclones, gas scrubber, fans and the stack shown in Figure 1, is extremely complicated. Our objective was to make a computer analysis of the existing system and design a system satisfying the ventilation needs, as at present the exhaust is not uniform. In addition, several new pots had been added, which led to some operational problems. As regards the exhaust system, on-site measurements were carried out first, and then a computational model was developed. The measurements were carried out with the aim of analysing the existing system on the one hand, and on the other, of determining the flow losses in the elements (see Schifter et al 1999).

The characteristic curve of the fans was also determined. On the basis of the findings, a numerical model and code were developed. The cooling of the hot gases as they proceeded through the exhaust system was taken into account in the design of the model (Czibere et al. 2000). Using the computer code, we evaluated various operational modes of the present system and discussed the improvement potential and possible changes required to improve efficiency. It was determined that there were ways of improving the system, but in order to achieve a truly efficient and uniform exhaust process, the whole system would have to be rebuilt and most pipes replaced. With the help of the code, the system has already been redesigned, and rebuilding is in progress.

Increasing the efficiency of the fan system

The computer simulation discussed above revealed that raising the efficiency of the exhaust system cannot be achieved without improving the fan capacity. For this reason it was decided to test the operational mode of the fan system by conducting research and making calculations, on the basis of which further tasks were determined. The activities carried out involved two main areas (see Szabó et al. 2000).

The first step was to determine the operational parameters of certain fans. This was followed by an investigation of the possibilities for increasing the capacities of the fans, which involved an analysis of ways of increasing the fan capacity without installing completely new fans in the system. The fan section consists of 4 identical radial flow fans which operate parallel to one another. A set of guide-vanes for prewhirl control is mounted at the inlet of the fans and a louver type valve is fixed at the outlet of the fan. The operation of the fan system was checked several times. In order to determine the characteristic curves of the fans, standard measuring conditions were fixed. The fan total pressure against volume flow rate was determined at various positions of the guide-vane regulator.

The determined characteristic curve was of help in the analysis of the current condition of the exhaust system and the identification of the expected operating point of the redesigned system. The measured results justified the need for an analysis of the increase of the operational capacity of the fans. This was why methods aimed at increasing fan capacity were investigated. It was concluded that an approximate 25% increase in capacity could be guaranteed without having to replace the electrical engines running the fans. Having evaluated all the options for increasing the capacity, three were chosen which could be applied with the built-in fans. These were as follows: increasing the speed of revolution by installing a drive mechanism between the engine and the fan; a slight increase in the size of the impeller; and finally, mounting a fifth fan parallel to the other four.

Having analysed the increased speed of revolution it was concluded that the number of revolutions could be raised from 1500 rpm to 1616 rpm. After consideration of the geometrical constraints, it was found that the size of the impeller could also be increased, in the manner shown in Figure 4, without changing the outlet blade angle. The installation of a fifth fan was also examined and the expected characteristic curve in all three cases was determined by applying similarity laws. Figure 5 compares the effects of the above-mentioned options where the guide-vane regulator is completely open, showing the extent to which they affect the original characteristic curve. As can be seen in the figure, all three variations alter the curve in the same way. The ideal solution to the problem would be to apply the method of increasing the speed of rotation, as this would result in a considerable increase in fan total pressure. The application of a fifth fan would marginally improve the upper part of the characteristic curve, where the operating point can be expected. A slight increase in the size of the impeller seems to have a favourable impact. On the basis of the calculations made, the expected costs can also be computed and it was determined that the most economical solution is to apply a completely new impeller of a larger size. This should not pose a problem as impellers always have to be replaced on a regular basis.

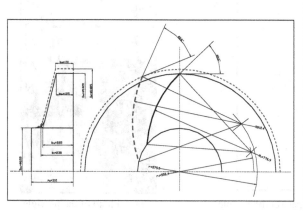

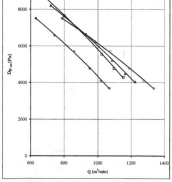

Fig.4. Increasing the size of the fan-impeller

Fig. 5. Alternative means of increasing fan capacity

Intensification of the gas scrubber

The equipment installed in the exhaust system and used for gas scrubbing plays an essential role in environmental protection. The hazardous gases flowing through the system are absorbed by the added alum. The alum is separated again and only clean air containing harmless gases is released into the environment through fans and the chimney. This is the reason why the gas scrubber is of such great importance. However, the operation of the gas scrubber caused some problems. The added alum agglomerated in the form of lumps and was not evenly distributed in the gas stream. This had several unfavourable effects. Firstly, the dust lumps eroded the walls of the channel, which is not streamlined from the point of view of fluid mechanics, resulting in heavy wear and holes in the walls. Secondly, certain sectors of the sack-type dust separator installed at the end of the gas scrubber became overloaded, whereas others hardly played any role in the separation. The most harmful effect was caused by the fact that the absorbing capacity of the dust lumps was considerably lower than that of the alum spread evenly in the mixture of air and gas to be cleaned. Thus, the gas scrubbing procedure was not ideal.

In order to find a solution to this problem, on-site measurements were carried out and the operating condition of the gas scrubber was checked so as to determine the flow losses in the channel. This was followed by modelling in the laboratory. The 1:10 scale model of the gas cleaner was built of plexiglas. Figure 6 shows a schematic drawing and Figure 7 a photo of the plexiglas model from another aspect. Pressure and velocity were measured in the model so that the flow conditions of the gas could be determined. This was visualised by the usage of a smoke band. On the basis of model tests, stagnation domains (wake) of the flow were identified, that is the spaces where considerable flow separation or other phenomena causing great losses were observed. Flow deflectors were installed in an attempt to bridge these problems, channelling the flow and reducing its resistance. After the suggested flow deflectors were installed, the resistance of the channel decreased by about 5%.

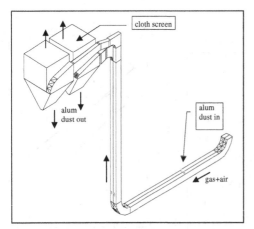

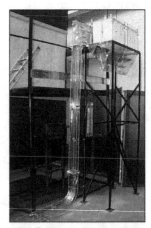

Fig.6: Schematic drawing of the gas scrubber's plexiglas model

Fig.7: Photo of the plexiglas model of the gas scrubber

The other weak point of the channel was the fact that the alum fed through primary and secondary dust inlets agglomerated due to gas flow, and proceeded through the system in the form of lumps. The places where these lumps came into contact with the wall of the channel underwent severe erosion. The agglomerated dust lumps also reduce the absorption capacity of the alum. This is why it is desirable to eliminate or inhibit lump development. One way of doing this is to install vortex generators in the flow. Vortex generators made of round-shaped plates were installed in several places in the laboratory model and their size, angle of setting and location were measured and optimised by conducting a series of tests. As a result, the risk of dust lumps forming was reduced to a minimum. However, the application of vortex generators always has an unfavourable effect. In our case, it meant that the flow resistance of the test section increased by 14.4%. Taking into account the whole exhaust system, however, the increase in resistance does not exceed 8%.

Figure 8 illustrates the application of flow deflectors and vortex generators installed in the bottom section of the gas scrubber, while Figure 9 shows them in the upper section. The photos in Figure 10 illustrate the flow without and with flow deflectors in the elbow of the gas scrubber. The flow without deflector, marked with oil mist, fails to follow the interior curvature of the elbow, drifting instead to the opposite wall and causing severe wear. When a flow deflector is used, the flow follows the arch.

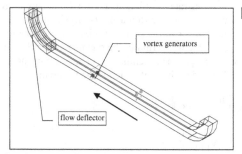

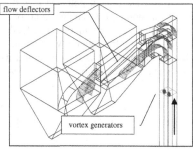

Fig 8: Application of flow deflector and vortex generators built in the bottom section of the gas scrubber

Fig. 9: Application of flow deflectors and vortex generators installed in the elbow of the gas scrubber

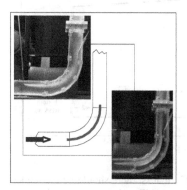

Fig. 10: Flow-visualization with and without flow deflectors in the elbow of the gas scrubber

After flow deflectors and vortex generators had been installed, wear decreased considerably, the gas cleaning efficiency of the alum increased significantly and the amount of alum used could be reduced without an unfavourable effect on cleaning efficiency. Thus, the early results are positive. The life span of the sack-type dust separators is likely to increase because of the even loading achieved.

Natural Ventlation of a smelter potroom

The ventilation of large potrooms, and of smelter potrooms in particular, (in this case, the measurements were: width 40m, length 800m, height 18m) can only be achieved by the use of natural ventilation. Ventilation is used to dilute and evacuate the emitted dust and gases from the buildings and to improve the working conditions of the employees.

A smelter potroom can be considered an open thermodynamic system. This system is not adiabatic, as there is heat transfer between the environment and the building. From the hot surfaces in the potroom, heat is emitted into the air of the potroom through the processes of radiation and convection. This is considered to be an internal heat source. Heat emanated from lighting, pot burners, pots and the heat of external surfaces of exhaust pipes entering the building heated by solar rays are also classed as internal sources. Natural ventilation is achieved by the hydrostatic pressure difference caused by the difference in indoor and outdoor temperatures.

Let us consider a typical potroom in order to demonstrate ventilation. Figure 11 shows a cross-sectional elevation of a smelter potroom. Due to pressure difference, the fresh air enters the potroom through a side window. The fresh air and the air in the potroom mix, and the air mixture leaves the building through the window in the roof. Due to continuity the quantity of air flowing in and out of the building is the same and its value is q_v. The ventilation of a potroom having a volume V is characterised by air exchange/hour L, the definition of which is the following:

$$L\left[\frac{1}{h}\right] = \frac{q_v\left[\frac{m^3}{s}\right] \cdot 3600\left[\frac{s}{h}\right]}{V\left[m^3\right]}$$

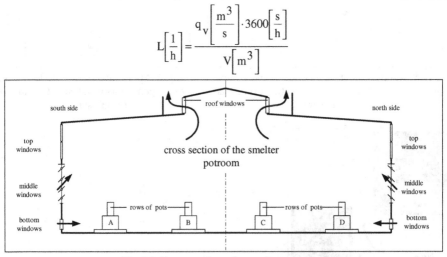

Fig. 11. Cross-section of a smelter potroom.

The desired air exchange number in the potroom is about 30/h; that is, there should be a complete exchange of air within the potroom 30 times each hour. As the propelling force of the air is caused by the difference in internal and external temperatures, it is clear that weather conditions considerably influence the ventilation. It is obvious that the airflow and the wind conditions also have a strong effect on ventilation in the potroom (Holt et al. 1999). Ventilation of windward and leeward areas differs. Ventilation of a potroom is also affected by openings such as windows. Whether the windows are open or closed, and the angles at which they are set are also important parameters of ventilation. Realising that the air exchange number depends on these factors, will provide a basis for choosing optimal ventilation conditions, with special emphasis on the weather conditions.

It is obvious that simply knowing the air exchange number does not provide enough data to characterise air exchange because, although this figure may represent an optimal rate, the flow can fail to mix the fresh air with the polluted air. Another problem arises when the ventilating airflow stirs up alum dust at the edge of the pots and keeps it in motion in the air of the potroom. Summarising the above-mentioned findings, we found that the analysis carried out regarding the natural ventilation of the potroom required a multi-angled approach. Thus, the investigation was carried out in various ways:

• A global computing method was developed to calculate the air exchange number. This process can be applied in the calculation of both indoor and outdoor temperatures and the computation of the air exchange number for different set angles of windows. The calculations were in line with the on-site measurements.

• Measurements aimed at determining the air exchange number were carried out under various weather conditions. The velocity distribution of the outflow of the air-gas mixture was measured along the upper windows in the potroom. There were several hundred measuring points. After evaluation of the results it was possible to calculate not only the air exchange number but also the distribution of the volume of the air flowing out of the potroom. The difference between the ventilation on both sides of the potroom was also observed.

• The flow pattern was investigated by applying fog candles, and a video camera was used to record the flow. Some shots are presented in Figure 12. Unfortunately, neither the video film nor the shots chosen from it are clear enough for a detailed investigation of the flow. The reason for this is that the video cameras had to be placed relatively far from the fog candles because of the strong magnetic field in the potroom. Secondly, three-dimensional flow cannot be illustrated in two dimensions. Nevertheless, this test was essential, as characteristic flow patterns were clearly visible to the eye. Such patterns were, for instance, how the fresh incoming air flows through the window, how strong the vortices are between the pots, what movement the gas makes before leaving the potroom through the upper windows, etc.

Fig. 12: Flow-visualization applying colour fog candles and a video camera

• Computer simulation was used to observe the distribution of velocity and pressure in the potroom. The commercial software package FLUENT has been successfully used by other research centres (e.g. University of Pretoria) to simulate three-dimensional flow inside and around buildings. We have also begun similar tests and the results are promising. The calculated results alone are of no use if there is no opportunity to check them with on-site measurements. Measurements on site are very important, but their number is limited because of the huge areas involved. This makes it essential to carry out tests on models.

• Model tests are very useful in situations where on-site measurements involve numerous and/or expensive measurements or difficult measuring conditions. In our case, the number of possible variations is unlimited as far as weather conditions and angles of the windows are concerned. The size of the building far exceeds the capacity for on-site measurements; thus, model experiments have been in the past, and still are now, an important aspect of our work. A 1:80 scale model of a part of the building was prepared.

Before placing the model in a climate-controlled wind tunnel a completely new measurements section was built by a member of the project team. The left-hand photo in Figure 13 illustrates the thermo-insulated wind tunnel; the photo in the middle shows an insertion with a flow direction from right to left; and the insertion can be seen placed in the upper part of the wind tunnel in the right-hand photo. As the wind tunnel is equipped with an air-conditioning system, it can be cooled or heated, and the humidity can be controlled. The wind tunnel is controlled by a PLC connected to a computer. The temperature and velocity distributions of the flow upstream of the model can be controlled. The model of the smelter potroom was placed in this part of the wind tunnel. A computer graphic of this model in different phases of its construction can be seen in Figure 14. The photo on the left illustrates the rows of pot models. They can be heated, and their structure is illustrated in Figure 15. Figure 16 illustrates the plexiglas model of the potroom, placed on the circular table of the wind tunnel. Both the model and the table can be placed in the flow at various angles and the windows of the model can be set at various positions. In order to visualise the flow inside the model under various weather conditions, oil fog was fed through the floor openings of the model.

A laser plane was developed to enable us to measure accentuation of the flow pattern emerging in the cross-sectional flow. The laser plane was created by a polygonal mirror rotating in front of a laser source of 25mW. Figure 17 shows the device with its front panel removed. Figure 18 shows a flow pattern at a certain window position, wind direction and external temperature. With the help of such photos from the wind tunnel model tests, it is possible to conduct the survey of flow patterns in the potroom.

Fig. 13: The thermo-insulated wind tunnel, the confuser insertion and the insertion placed in the upper part of the wind tunnel

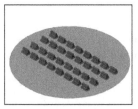 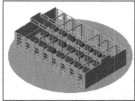 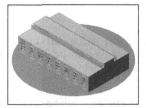

Fig. 14: Computer graphic of the potroom model in different phases of its construction

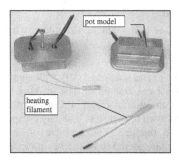

pot model

heating
filament

Fig. 15: The pot model

Fig. 16: The potroom model made of plexiglas
placed on the circular table of the wind tunnel

PSU

polygonal
mirror

laser

Fig. 17: Equipment for making a laser plane

Fig. 18: A flow pattern in the potroom model at a
certain window position, wind direction and external
temperature

208

Conclusions

In order to achieve the desired ventilation, to improve the comfort of the workers and to minimise the emission of pollutants into the external environment, ventilation should be adjusted to various weather conditions and:

- the air exchange number should be 30/h;
- the temperature in the area used by workers should be neither too high, nor too low;
- the draught of air on the working floor should be acceptable;
- conditions in the sections of the potroom which are not close to the windows should also be satisfactory.

The open part of the pots is situated at a height of about 0.5m to 1 m and is covered only with alum. The earth is regularly loosened by machines, which generates a large amount of dust, making it necessary to minimise the air flow at this level so that the dust can settle. A cloud of dust will not settle in a strong airflow, but instead will pollute the air in the potroom or, worse still, it may escape from the potroom at the roof, causing severe environmental damage. The problem of the sedimentation of the dust both inside and outside the potroom is considered to be the key issue in the reduction of pollution

The main objective of this study is to solve these complex tasks associated with ventilation. Finally, it can be concluded that the increase of the capacity of the exhaust system, the increased efficiency of gas and dust separation and the proper control of ventilation of the potroom will result in:

- the improvement of internal air conditions in the potroom;
- the reduction of environmental pollution by emissions from the potroom.

The results of the conducted research and the methods applied can be used to reduce pollution from industrial potrooms and the environmental harm caused by other technological processes. The method of investigation can be further enhanced by the combination of numerical simulation and optimisation. In this particular case, the software package FLUENT used in the simulation of interior and exterior flow of the potroom would be extended by an iterational algorithm. In this formulated target function, for example, we are attempting to identify the maximum air exchange number under various weather conditions. The variables include the angles at which ventilation windows are set. This algorithm provides information concerning the position of the windows so as to achieve maximum ventilation, taking into consideration the specified weather conditions This algorithm will be developed in co-operation with experts from Pretoria University within the framework of a Hungarian and South African exchange programme. Figure 19 is a graphic from the first simulation carried out by the FLUENT programme system. It shows the distribution of temperature in a two-dimensional potroom model, in still air (left) and with a side wind blowing from the left (right). Further investigation of this type should provide an even more accurate understanding of flow and ventilation of potrooms and other industrial facilities.

209

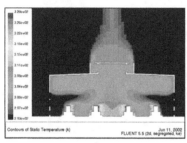 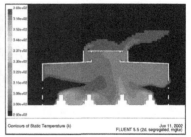

Fig.19: Temperature distribution in a potroom model, in still air (left) and with a side-wind blowing from the left (right)

References

Czibere, T., Kalmár, L and Pap, E. (2000) Numerical simulation of compressible fluid flow in pipeline systems, Proceedings of Micro CAD International Computer Science Conference, Miskolc, Hungary, 2000. Session M, pp. 37-44.

Holt, N.J., Anderson, N.M., Karlsen, M. and Foosnaes, T. (1999) Ventilation of potrooms in aluminium production. Light Metals 1999, The CD-ROM Collection, The Minerals, Metals & Materials Society.

Schifter F., Szabó, Sz., Czibere, T. and Kalmár, L. (1999) Investigation of Gas Exhauster-system in a Soederberg Plant, Proceedings of Micro CAD International Computer Science Conference, Miskolc, Hungary, 1999. Session M, pp. 141-146.

Szabó Sz., Schifter, F., Juhász, A. and Baranyi, L. (1999) Exploration of Functional Disorders of Gas Exhauster System in a Soederberg Plant, 11th Conference on Fluid and Heat Machinery and Equipment, Budapest, 1999, on CD Rom, 1-6.

Szabó, Sz. et al., (2000) 'Kohócsarnoki gázelszívó rendszer felülvizsgálata és áttervezése (Analysis and redesign of gas removal system of a smelter potroom)', Research Report, 2000, Department of Fluid and Heat Engineering, University of Miskolc, vol. 5.

CHAPTER 22

Environmental Effects of Ferrous Slags
- Comparative Analyses and a Systems Approach in Slag Impact Assessment for Terrestrial and Aquatic Ecosystems

Dr. Lidija Svirenko, Jurij Vergeles, Oleksandr Spirin

Department of Urban Environmental Management & Engineering (UEME),
Kharkiv State Academy of Municipal Economy (KSAME),
vul. Revoljuciji, 12
UA-61 002 Kharkov
Ukraine

Summary

On the basis of laboratory and field studies of properties of Martin, converter and blast-furnace slags from iron works in the Ukraine, Russia and Latvia it has been found that their processing, storing and use in construction cause significant changes in the environment, particularly, in the chemistry of surface and ground waters, atmosphere, soils and rocks. Shifts in the environmental conditions in the areas affected by ferrous slags, while simultaneously affected by other human-caused factors, leads to qualitative and quantitative deterioration in the composition and functioning of aquatic and terrestrial biocoenoses. To prevent or mitigate such negative trends a systems approach and some environmental and ecological technologies (e.g. utilising ferrous slag in construction, together with coal industry solid wastes, establishing constructed wetlands for treatment of wastewater from metallurgy works etc.) are proposed for dealing with ferrous metallurgy and rock mining wastes.

Introduction

Slag is an unavoidable by-product of various recasts in the metallurgical cycle. The output of slag can reach from 10 to as much as 40% in the production of irons. Most blast furnace slag is reprocessed into granulated slag for the production of cement, porous concrete aggregates and crushed stone. Besides, Martin, converter and blast furnace slags are used in metallurgy, production of fertilisers, soil conditioners and abrasive materials. However, in the former USSR the proportion of utilised steel slag did not exceed 23 - 30% since most slag was merely dumped.

Most types of slag are used as rough crushed stone or fractionated gravel in hydraulic, road and civil building and land planning. The use of slag in building in place of rock materials means a reduction in the workload compared with quarrying, and at the same time reduces the volume and area of slag heaps at most metallurgical plants in Eastern Europe.

Slag crushed stone is produced by quick cooling of melted slag or by working slag heaps, where slag may age or have been subject to weathering for dozens of years. Various types of slag are widely used in the industrial water supply and construction of slurry sewer works (industrial waterworks) at plants and mines in the Ukraine, in bank protection structures and other elements of hydraulic works at plants in the Ukraine, Russia, Germany etc (Svirenko, 1976; Svirenko et

211

W. Leal Filho and I. Butorina (eds.),
Approaches to Handling Environmental Problems in the Mining and Metallurgical Regions, 211–229.
© 2003 *Kluwer Academic Publishers. Printed in the Netherlands.*

212

al.,1983; Olpinski & Christensen, 1981; Verma, 1982; Brjeslavjec et al., 1985). These have shown how efficient slag is as a construction material. This is connected primarily with the thickness of slag crushed stone.

At the same time slags represent multi-component systems consisting of high-temperature minerals - products of special technologies - and are unstable under when they come into contact with the ground. When they interact with moisture and atmospheric gases, secondary minerals are produced, i.e. the process of weathering occurs. This phenomenon is partly responsible for such peculiarities of slag as a certain disintegration with time and the existence of hydraulic activity (self-cementation).

It is reasonable that during the process of slag material production the characteristics of the water and air which come into contact with the slag, change (Abramovich, 1977; Andon'jev, 1979; Futamura, 1981; Kormyshjev et al. 1981; Shkol'nik et al., 1983; Chudakov & Tihonova, 1985). However, the literature available provides little information on the impact of slag on the chemistry of water bodies and the consequent shifts in the living conditions of aquatic organisms, and the information given often deals with isolated aspects and is not systematised (Bronfman, 1972; Bronfman & Hljebnikov, 1985; Vinogradov, 1991). The aim of this article is to show the whole spectrum of slag impacts on ecosystems and the urgent need to introduce special environmental and ecological technologies to prevent pollution by slag components when treating slag treating and using it in civil engineering (Verma et al., 1982; Sehi et al., 1986; Svirenko et al., 1990).

Geography of Investigations

Investigations are being carried out at the following metallurgical plants in the Ukraine: *plants of the middle Dnieper:* KryvorizhStal Combine Works, ZaporizhStal Combine Works [ZSCW], 'F.E.Dzerzhinskij' Iron Works [DIW]in Dniprodzerzhinsk; *plants in the Donjec Basin:* 'Illich' Iron Works, AzovStal Combine Works [ASCW], Makijivskij Iron Works, Jenakijivskij Iron Works, Alchevskij Iron Works), Russia (Chjerjepovjeckij Combine Works, 'Krasnyj Oktjabr' Iron Works in Volgograd) and Latvia (Liepaja Metallurgical Plant [LMP].

Materials and Methods

We examined blast furnace, converter and Martin slags, both fresh and old slag in slag piles, as well as the surrounding area as affected by the treatment and use of the slags. The authors used a system approach in planning investigations and analysing results. It has been assumed that the introduction of slag into natural systems would cause shifts both in abiotic (physical characteristics, water chemistry and chemistry of air and soils) and biotic components of ecosystems (Kisiljevskij et al., 1980; Mjetodichjeskije, 1988; Lapchinskaja et al., 1988).

The chemical composition of various types of slag was ascertained by routine methods, and trace elements (Cu, Zn, Ni, Co, V and Pb) were determined with the aid of emission spectrum analyses. The nitrogen content in slag was determined with the aid of gaseous analyses in the Central Research Institute of Ferrous Metallurgy (CNIIChjerMjet, Moscow, Russia).

The mineralogical composition of slag was studied with the aid of crystal optics methods and roentgen-phase analyses. The water quality and chemistry of the filtrate were monitored by the following indices: pH, eH, dry residue, COD_{Mn}, alkalinity, hardness, PO_4, $N-NH_4$, $N-NO_2$, $N-NO_3$, Cl^-, SO_4, H_2S using standard methods. Concentrations of metals (Fe, Mn, Cr etc.) were determined using an atomic absorption analyser.

Individual ecosystem processes involving the contact of slag with water were modelled under laboratory conditions using various types of water (river, sea and distilled water), varying process duration and flow rate of filtrate abstraction. The process was also evaluated under natural conditions.

Results

Slag heaps and the use of slag in construction , as factors in technogenic complexes, cannot be considered in isolation without considering their interaction with other components of natural ecological/geochemical systems. Diverse processes of substance exchange (water exchange, air circulation, biological migration etc.) provide its re-distribution between natural and technogenic systems, particularly in slag and surrounding elements.

Therefore, the main task of the research was to investigate changes in slag composition and objects which came into contact with it , and to examine the accumulation of trace elements in living organisms.

Slag composition

In terms of their chemical composition all types of slag examined belong to the category of basic slags. The highest basecity was found in Martin phosphorous slag. A wide spectrum of trace elements was also determined in slag. Concentrations of trace metals are shown in Table 1.

Table 1.Trace metals in various types of slag, mg/kg

Type of slag	Quantity of samples	V	Cr	Ni	Cu	Zn	Pb	Ti
Blast furnace slag	6	20-110	8-105	no data	10-30	80	8-12	140-1000
in average		42	32		16	80	8	456
Martin and converter slag	9	58-940	1850-9580	9-68	19-54	63-320	0-20	371-510
in average		229	4487	30	31	103	7	441

Such variations in metal content depended on the composition of the charge. The following variations in nitrogen content were also found, which agree well with existing literature (Fix et al., 1972; Table 2).

Table 2. Nitrogen content in slag

Type of slag	Place of origin	N content, %
Blast furnace slag	ZSCW	0,02701
Blast furnace slag	ZSCW	0,03445
Blast furnace slag	ZSCW	0,02844
fresh Martin slag	LMP	0,03570
fresh Martin slag	LMP	0,01245
weathered Martin slag	LMP; weathered under natural conditions	0,06316
weathered Martin slag	LMP; weathered under experimental conditions	0,04061
weathered Martin slag	LMP; weathered under experimental conditions	0,04169

Mineralogical composition

Blast furnace, converter and Martin slags differ significantly in their mineralogical composition. Weathered slag can be distinguished from fresh slag by the presence of a thin layer of secondary minerals.

The most prevalent minerals in slag are silicates. In blast furnace slag, silicates are represented by a series of solid solutions such as melenites $Ca_2MgSi_2O_7$ - $Ca_2AlSiAlO_7$, and calcium silicates - larnite, rankinite, and pseudowollastonite.

Larine $Ca2SiO4$, silicates of calcium and magnesium such as mervinite and monticellite containing tricalcium silicate prevail in Martin and converter slags.

In the case of a high phosphorus content in slag, phosphorus-containing minerals, such as nagelschmidttite, $Ca_3(PO_4)_2 \cdot 2Ca_2SiO_4$, are formed; in the opposite case the phosphate group is included in bicalcium silicate.

Other characteristic minerals in slags are braunmillerite, $4CaO \cdot Al_2O_3 \cdot Fe_2O_3$ and mono- and bicalcium ferrite. The presence of spinellides of various composition ($FeCrO_4$, $MgFe_2O_4$, $MnAl_2O_4$ and $CaCr_2O_4$), of Fe and Mn oxides, CaS (point inclusions in glassy mass) and quicklime CaO was observed. Nitrogen, according to Fix (1975), is present in slag both in dissolved form and as metal nitrides, particularly like easily hydrolysed Ca_3N_2 (Litvinova et al., 1972; Svirenko, 1978; Sajdullin, 1979; Pjerjepilicin, 1987).

Micro-porphyric structures are characteristic of all types of slag examined. These structures are formed by combining glassy phase with crystalline settlings of minerals.

Secondary minerals

These develop on the surface and into slag fissures. Calcium carbonate, $CaCO_3$, in the form of faterite and calcite, sometimes with magnesite, $MgCO_3$, and also ankerite, $Ca(Mg, Fe)(CO_3)_2$, siderite, $FeCO_3$, quartz and chalcedony, SiO_2, gypsum, $CaSO_4 \cdot 2H_2O$, and calcium hydrosilicates are common minerals. Bemite, AlOOH, and goethite, $HFeO_2$ are also present. Brusite, $MgOH_2$, is formed under interaction of slag with sea water. We observed phosphates, bobierite $Mg_3(PO_4)_2 \cdot 8H_2O$, brushite $CaHPO_4 \cdot 2H_2O$ and whitlockite $Ca_3(PO_4)_2$, in association with calcium carbonate in phosphorous slag (Svirenko, 1978).

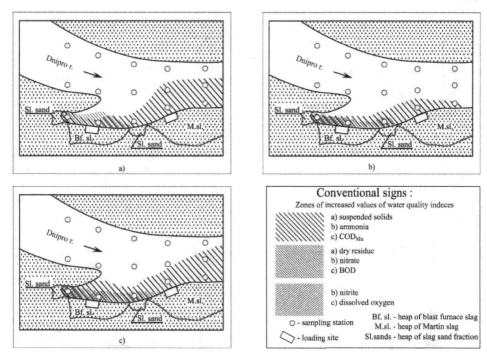

Fig. 1. Pollution of the Dnipro river from the premises of slag treatment yards and loading sites near DIW, after Stankovych (1983) with modifications

Changes in temperature and shifts in aquatic chemistry

When water and slag interact, changes in temperature (e.g., off ASCW and LMP) and significant shifts in aquatic chemistry are observed both in laboratory and during in situ measurements near iron works and hydraulic slag works. The most pronounced changes are an increase in pH, decrease in Eh and depletion of dissolved oxygen (especially in Dnipro water near DIW), an increase in suspended solids, salt content, values of BOD and COD, and trace elements, regardless of water type (Stankovych, 1983). The appearance of sulphides, nitrogen compounds and phosphates in solutions should also be noted.

216

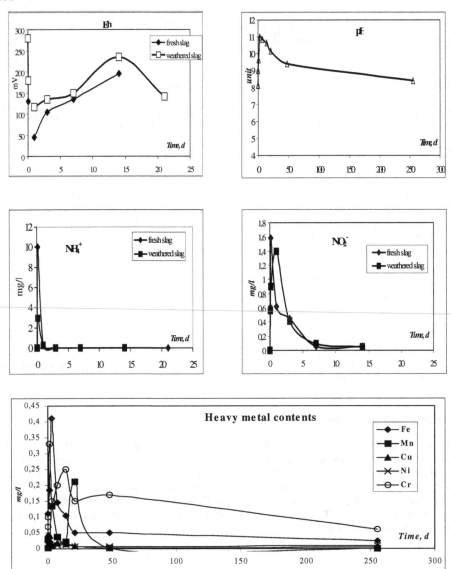

Fig. 2. Changes in Eh, pH, nitrogen compounds and heavy metals when water filtered through fresh and weathered
Martin slag in lab flush experiments with samples from LMP slagheaps
(Initial water: Eh = 280 *mv*; pH = 8.15; flow rate was 1 *l* of water *per* 1 kg of slag *per* day)
a) Eh; b) pH; c) NH$_4$; d) NO$_2$; e) heavy metals

In the course of flushing experiments drastic changes in all indices of aquatic chemistry were
peculiar to the first day of the experiment only. Then, concentrations of all components gradually
decreased and towards the end of the experiment, depending on the flushing rate, stabilised
approaching the indices of the initial water.

A near-balanced state was achieved more quickly when water interacted with weathered slag. In this case, concentrations of pollutants migrating into the water in the early stage of interaction was lower than in the case of fresh slag.

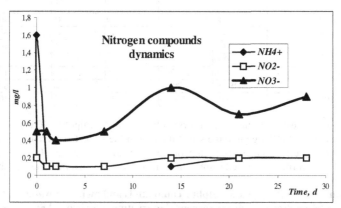

Fig. 3. Changes in ammonia concentration when water filtered through blast furnace slag in lab flush experiments with samples from ZSCW slagheaps (flow rate was 0.54 *l* of water *per* 1 kg of slag *per* day)

Water quality indices differed significantly from background ones in the Sea of Azov off the area of fresh blast furnace, converter and Martin slags fill, forming a sea dam, were the following (Table 3).

Table 3. Sea water quality in the course of slag dam building, ASCW

Sampling site	Water quality indices, mg/m³								
	pH	COD_{Mn}	$N-NH_4$	$N-NO_2$	$N-NO_3$	S⁻	SO_4	Mn	Cr
Near slag dam	10.15	14.88	0.32	0.012	not found	0.07	812.5	0.054	0.075
Ljapino (backgrd conc.)	8.85	6.4	0.04	0.00	0.32	0.015	725.0	0.031	Not found

Such shifts in water quality, including an unavoidable increase in suspended solids, created toxic conditions for living organisms. Thus, the occasional fill of several tons of slag into a concrete groove, modelling the system of a stretch of the Dnipro-Donbas canal (near the Orel'ka river) without flushing conditions, resulted in a dramatic increase in pH values, up to 14 in two hours, and, as a result, death of numerous aquatic organisms, including all molluscs, rotifers, crustaceans, amphibians and insects.

At the same time, results of long-term water quality monitoring in the Sea of Azov 100 m from the slag dam, built more than 30 years ago, showed that average values of the main indices were as follows: pH - 7.16, dissolved oxygen concentration - 6.83 mg/l, NH_4 – 0.453 mg/l, PO_4 – 0.18 mg/l (Bryginec et al., 1999). It has been observed that in industrial freshwater storage ponds of long-standing exploitation with dams and slopes stabilised with slag crushed stone water, quality indices did not differ significantly from background values (iron works of Donbas, LMP and the Liepajas Lake, Latvia etc.). A considerable accumulation of metals in <u>sediments</u> of bodies of water close to iron works was observed. Thus, in the northern part of the Liepajas Lake near the LMP, concentrations of Cr, Ni, Mn, and Fe in sediments showed a 3 to 13-fold increase compared with its central part of this lake (Liepa et al., 1989; Svirenko, Spirin, 1995; Svirenko, Yurchenko, 1995).

218

Table 4. Metal content in sediments, the Liepajas Lake, Latvia

Sample point	Fe, mg/kg	Mn, mg/kg	Cu, mg/kg	Ni, mg/kg	Cr, mg/kg	Zn, mg/kg
1 & 3	30500	1700	30	35	130	70
2	25000	1900	40	46	80	60
4	13000	2000	30	8	10	80
5	9000	600	20	8	10	80
7	13000	1700	15	8	10	80
8 & 9	38000	3800	55	34	63	80
10	9000	520	20	8	10	80

Ground water was investigated in the area formed by slag in Mariupol (ASCW) and Zaporizhzhja (ZSCW). Groundwater discharge into the sea at the foot of the slope from site of the slag treatment division (ASCW) resulted in increased temperature, pH values exceeding 10, high content of H_2S, sulphides, and other forms of reduced sulphur, and the presence of several metals (Table 5).

Table 5. Contents of nitrogen and sulphur compounds and metals in water and wastewater samples taken at the ferrous slag treatment division, ASCW

Kind of sampled water	T, °C	pH	NH_4^+, mg/l	$H_2S + HS^- + S^{2-}$, mg/l	$S_2O_3^{2-}$, mg/l	SO_3^{2-}, mg/l	Fe, mg/l	Mn, mg/l	Cr, mg/l	Co, mg/l	Ni, mg/l	Zn, mg/l	Cd, mg/l
Piped water from process water supply system		7,40	0,36	7,24	56,05	2,40			not found				
Wastewater from gravel-making of converter slag		8,15		6,40	15,70	12,10	0,21		not found	0,25	0,75	0,28	0,031
Wastewater from punice-stone production	32	11,1		352,00	1792,0	768,60	0,18	0,04	not found	0,38	1,40	0,04	0,046
Wastewater from granulated slag production	60	9,50		51,20	112,05	62,40	0,32	0,04	not found	0,25	0,75	0,01	0,031
Wastewater from gravel-making of blast furnace slag	36	10,75			1096,00	1708,6	430,00	0,12	not found	0,25	0,60	0,05	0,046
Groundwater seepage to the buffer pond	40	10,47			1174,40	560,50	272,00	0,09	not found	0,19	0,40	0,02	0,046
Groundwater seepage to the Sea of Azov	45	10,45			851,20	399,10	57,60	0,11	not found	0,19	0,75	0,06	0,031
Sea water (control)	21	7,40	2,3	7,68	3,36	27,20	0,48	0,08	not found	0,19	0,60	1,19	0,031

Water in the wells on the slag treatment sites (ZSCW) was distinctly metamorphized. Its salt content was as high as 3000 mg/l or more, and the pH value reached 12. It was sulphate-rich as regards its anionic composition.

Changes in ground features under the influence of thermal mineralised water

The basement of the manufacture zone in ZSCW near the slag treatment yard turned out to be formed mainly of slag. The effect of thermal water on the slag massif on the site of the blast furnace slag treatment plant for the production of crushed stone (ZSCW) produced the following:

- temperature in wells at a depth between 1.5 and 30 m was as high as 76°C;
- the water is calcium-magnesium-sulphate-rich;

- the emission of gases such as H_2S, SO_2, H_2SO_4, NH_3 and CO has been observed at the mouths of the wells;
- a characteristic feature of slag from samples taken at depth was the significant transformations, such as a decrease in solidity and thickness, widespread development of zeolites, a calcium series (in addition to calcium hydrosilicate and sulphate). These minerals are peculiar to natural hydrotherms.

Impact on atmospheric environment

In experiments with blast furnace slag under exposure to water ($t = 25°C$; ionic force, $I = 0.025$) emission of hydrogen sulphide was determined: quantities calculated quantities amounted to 671 mg/m^3 in 1 hour, 55 mg/m^3 in 1 day, and 7.5 mg/m^3 in 14 days. Emission of ammonia from slag was observed during the first week only. At the same time, concentrations of H_2S (up to 130 mg/m^3), SO_2 (up to 261 mg/m^3), NH_4 (up to 208 mg/m^3), H_2SO_4 (up to 1.46 mg/m^3), C_2H_2 (0.23 mg/m^3), CO (up to 5 mg/m^3) were found at the mouths of wells drilled near the slag treatment yard, where slag crushed stone is produced by watering hot slag. The temperature at the mouths of the wells varied between 23 and 76°C.

It is important to note that simultaneously with gas emissions, solid particles entered the atmosphere both during metal smelting and melted slag treatment.

Soils of industrial and sanitary protective zones of ASCW, where slagheaps were situated from the 1970s to the 1990s, contained quantities of metals, such as Fe, V, and Mn, 2 to 3 times higher than soils in the residential area of Mariupol. This might be connected not only with precipitation of solid particles of slag but also with dust from vent systems of the main shops. The high metal content (Fe – c. 10%; Zn, Cr and Pb – more than 0.1%) in the sedimentation pond at the gas and dust-cleaning unit of LMP confirmed this possibility.

Impact on biota

An increased metal content was observed in the biomass of plants growing on slag substrates or near slagheaps. Thus, metals such as Ni, Zn, Ag had accumulated in leaves of the poplar (_PopulusX sp._) and robinia (_Robinia pseudoacacia_), and, in addition to these three metals, Cu and Ba, were also found in grass cover near ASCW. In Liepaja, herbs growing on slag dams were enriched with Fe and Ni. Under general high content of Fe in their ash, mosses on slag dams both recently filled and long- established, had a markedly high level of Zn and Pb (Fig. 4).

220

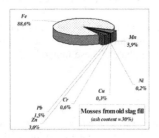
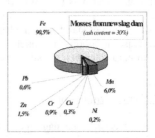

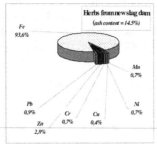

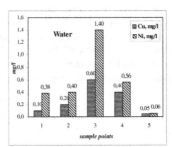

Fig. 4. Comparative metal contents in moss and herb ash from slag dams a) of long-standing fill; b,c) of recent fill, LMP

Cattail (*Typha latifolia*), a dominant helophyte species on the site of the wastewater treatment station of the LMP and by the lakeside, shows significant ability to accumulate metals in a polluted environment. Fe, Mn and Cu levels in cattail ash increased by factors from 3 to 15 (Figs. 5 and 6).

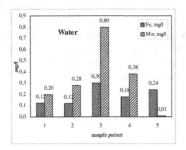

Fig. 5. Iron, manganese, copper and nickel contents in water from the industrial water supply system, LMP
Here and in the next figure: sampling point 1 – water abstraction intake from the Alande river; 2 – pond for water supply; 3 – wastewater discharge; 4 – wastewater after cattail stand; 5 – the Liepajas lake (middle part)

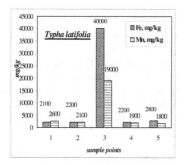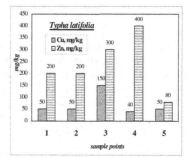

Fig. 6. Accumulation of iron, manganese, copper and zinc by cattail, _Typha latifolia_ (in ash), the Liepajas Lake, Latvia

The gradual reduction in the biodiversity (biodiversity index decreased from 1.7 to 1.3 over the 15-year period from 1976 to1991) of plankton and benthic communities and especially significant shifts in species composition of bottom fauna were observed on the section from the open sea towards the coastal part of ASCW, near the place where it discharges thermal sulphide waters. Thus, on the off-coast stations near ASCW the dominant species were _Mya renaria_ (_Mollusca_) and crab _Rithropanopeus harrisii tridentatus_ (_Decapoda: Crustacea_) these being new species for the Sea of Azov. Collections made in the stations towards the open sea showed partial enriching of benthic communities at the expense of such species as _Nereis sp._, _Hypolimella viola_ (_Polychaeta: Annelida_), _Pectinaria kovalewskii_, _Hydrobia talinassii_ (_Mollusca_), _Balanus improvisi_ (_Cirripedia: Crustacea_), and _Cerastoderma lamarcii_ (_Mollusca_) with low abundance (Britajev, 2002).

In the northern part of the eutrophic Liepajas Lake, adjoining LMP, dramatic decreases in species diversity, amount and biomass of plankton and bottom infusorians (important consumers of detritus and bacteria) and macrobenthos organisms were observed. Only the most common species of _Mollusca_, such as _Euglesa sp._, _Bithynia sp._, _Valvata sp._ and _Limnea sp._ were found. On the whole, numbers of _Bivalvia_ and _Gastropoda_ were significantly lower than in other parts of the lake. A leading role in the lake's benthos community was played by oligochets _Limnodrillus hoffmeisteri_ and _Potamothrix hammoniensis_ which accounted for more than 50% of all organisms in terms of density of population. Considerable changes in morphology and life forms in some aquatic organism groups have been revealed. These indicated a certain toxicity of the water, aggravated by the raised water temperature in the northern area. This again was confirmed by not infrequent findings of algal cells with an evident disintegration of chloroplasts, numerous empty algal cell envelopes, abnormal forms of crustaceans and a great deal of dead benthic organisms, particularly macro- and microphytobenthic algae and molluscs, up to 80%, or 2 to 3 times more than in other parts of the lake (Liepa et al., 1989; Zarubov et al., 1989).

Discussion

The role of pH and reducing conditions in water-sediment equilibrium. Data on metals appeared rather ambiguous: sediments are the media which deposit chemical elements entering water systems, both in a suspended and in a dissolved state. However, under forming reducing conditions on the interface "water-sediments", processes of desorption can dominate and, as result, substances from both organic and mineral pools of sediments will be released into the water (Bronfman, 1972). Nitrogen compounds and phosphates leaching from slag are factors intensifying eutrophication of water bodies (Liepa et al., 1989; Svirenko et al., 1990; Vinogradov, 1991; Svirenko & Spirin, 1995).

<u>Fresh and weathered slag.</u> The intensity of slag-water interaction fell gradually and with time, owing to the creation of a surface skin of newly formed minerals which are tolerant to water. Consequently, weathered slag (from heaps) when used for construction purposes, has a less intensive effect on the characteristics of the medium water. Data of gaseous analyses, according to which nitrogen concentration in weathered slag increased (Table 2), are questionable, since this feature is not characteristic of sedimentary minerals. A possible source of nitrogen in sedimentary formations on the base of slag may be products from microbiological activity. It is not excepting that these results may be artefacts of this method.

<u>Blast furnace and Martin slag.</u> The role of the chemical composition of slag, i.e. the dependability of technological peculiarities of iron production on specific metals appearing in filtrate is central. Thus, the characteristic features of blast furnace slag are a high sulphur content and lower basicity than steel slag. In contrast, steel slag has a high level of calcium, phosphorus (compared with blast furnace slag), and trace metals, depending on the grade of the produced steel. These define specific features of environmental pollution components in the course of slag weathering and treatment.

Results of laboratory experiments and in situ observations and measurements have shown that slags impact all areas of the environment. Nevertheless, aquatic chemistry shows dramatic changes. A wide spectrum of indices, like pH, Eh, dissolved oxygen, BOD and COD, nitrogen and sulphur compounds, phosphates, metal ions etc., are prone to significant shifts.

<u>Slag treatment technology</u> In Ukrainian iron works it has been found that a continuously acting negative factor affecting the environment is the functioning of slag treatment yards in their present form. High-temperature wastewater discharge saturated with compounds of reduced sulphur (sulphides) is typical of the treatment of blast furnace slag. Entering surface and ground waters, they increase in temperature and form oxygen deficits. According to our calculations for ASCW, 2000 mg/l of oxygen are required to oxidise sulphur compounds in ground water from the slag treatment yard. Concurrently, the same index for process water from the used slag treatment division was no more than 78 mg/l.

The leading role of thermal processes in blast furnace slag treatment was confirmed by the patterns of distribution of the spots with "anomalous" water quality on the Dnipro river stretch near slag loading sites (Fig. 1). For example, DIW has a quarry for the production of crushed stone from blast furnace slag on the left bank of the Dnipro river. The river water cools the free-running slag which has a temperature of about 1300°C . The crushed stone was moved to a loading site by bulldozer and then shipped to consumers. Increased concentrations of suspended solids, ammonium, nitrite and nitrate nitrogen were found in river water along the loading sites. The concentration of nitrogen compounds was 10 to 40 times higher than in the upstream section of river. An increase in BOD and COD and decreased concentrations of dissolved oxygen are also recorded on this stretch of the river.

It is evident that "extreme" spots are directly confined to the shoreline near the yards which crushing blast furnace slag ($t°_{av.} \approx 1300°C$). Waste waters with flinders of slag from these yards make their way in the river unhindered. Light particles of porous slag are carried great distances – many dozen kilometres from the yards - by the river or waves of the sea (Karjakin, 1949; Svirenko & Spirin, 1997).

<u>Toxicity and biotic shifts.</u> Increased temperature of the surroundings, increase in pH, oxygen deficit and a high content of suspended particles appear to be the main factors determining toxicity in

environmental impact zones of slag treatment and hydraulic work construction involving slag. This toxicity is responsible for the reduction in biodiversity in aquatic zoocoenoses, domination of one or two tolerant forms and proliferation of abnormal forms () observed in brackish waters both of the Liepajas Lake (LMP) and the Sea of Azov (ASCW). A characteristic feature of plants growing within environmental impact zones of iron works is metal accumulation, in the course of which they take in metals from the water and from bottom substrates (higher aquatic vegetation) or from the soil and atmosphere (mosses and arboreal plants). A close correlation between metal content in surface waters and in ash of plants (Table 6) was demonstrated.

Table 6. Correlation between metal content in lake water and cattail (in ash), the Liepajas Lake, Latvia

Sample point	Fe (water), mg/l	Fe (Typha), mg/kg	Mn (water), mg/l	Mn (Typha), mg/kg	Cu (water), mg/l	Cu (Typha), mg/kg
1	0,124	2100	0,200	2600	0,100	50
2	0,120	2200	0,280	2100	0,200	50
3	0,300	40000	0,800	19000	0,600	150
4	0,180	2200	0,380	1900	0,400	40
5	0,240	2800	0,010	1800	0,045	50
r(Pearson)=	0,784		0,888		0,753	

Conclusions

Slag is an inevitable by-product of metallurgical processes, therefore iron and combine works make efforts to have it processed. At present, several leading companies have achieved 100% slag processing (e.g. chapter by Still, elsewhere in this book).

At the same time, control services in smelting works of the fSU which carry out monitoring of gas emissions and wastewater effluents generated during slag processing do not provide full and unbiased information on the presence and input of numerous components which have a harmful effect on the environment. As is evident from the above, by-products of slag processing are a serious factor, which has a significant effect on the natural and socio-cultural complexes adjacent to any iron or combine work.

In order to reduce the harmful effects of slag disposal and processing we propose, following the structural and hierarchic approach to urbanised systems developed by the authors (Shatrovskij, Spirin, 2002), to develop and establish a system of interrelated measures, which could improve and contribute to the existing systems. Within the framework of this system we distinguish between technological, control and organisational / administrative components:

The Technological component includes the following measures:

1. It is vital that we prevent penetration of wastewater from slag treatment yards into subsurface waters, and provide collection of surface runoff *in situ* to protect water bodies from progressing pollution.
2. As the activity of slag in interaction with water gradually decreases in the course of time, it is worth exposing fresh slag to artificially accelerated weathering before using it in construction. This would lead to the formation of a protective 'skin' on the slag surface and could only be implemented in special yards equipped with waterproof screens and facilities for collecting the filtrate.

224

3. In regions with a well-developed coal mining and smelting structure it is helpful to combine ferrous slag, which is basic, with coal-concentrating wastes, acidic in nature, in civil and hydraulic engineering. This will allow plants to include coal-concentrating wastes from dumps and tailings in the process of waste utilisation (Svirenko et al., 1992).
4. The introduction of ecological technologies, such as 'constructed' wetlands, is strongly recommended for water treatment and conditioning in water recycling units at mining and smelting works (Svirenko, Spirin, 1997). On the one hand, this will reduce the content of nutrients, suspended solids and toxic metals in water, and on the other hand, will saturate it with dissolved oxygen (Jakubovskij, 1975; Hejný, Sytnik, 1993; Zolotuhin et al., 1995; Stolberg et al., 2002).
5. It is also advisable to set up 'green belts', planted with carefully selected trees and bushes, around metal works. This will help improve surface air quality in these areas (Tarabrin et al., 1984; SNiP, 1984).

The Control component includes the following measures:

1. Monitoring of a wide spectrum of chemical components in fusion mixture (input) and produced slag (output) is vital if metallurgical enterprises are to effectively control wastewater effluents and gas emissions.
2. It is necessary to introduce reliable control of the chemical composition and quantity of wastewater produced in slag processing. Special attention should be paid to compounds of sulphur and nitrogen. Concurrently, considerable importance should be attached to the maintenance of corresponding acid-base equilibrium in wastewater since this determines the transfer of numerous critical elements and compounds from insoluble to soluble forms, enriching wastewater and adding to the toxicity of wastewater.

Organisational / Administrative measures:

As far as the Ukraine is concerned, all the above mentioned reorganisations in the systems of waste control and slag processing are possible only after significant changes have been made in the decision-making process, from the shop/factory floor right up to governmental and ministerial level.

Old-fashioned modes of development stemming from the industrial period and inherited from Soviet times where heavy industry was prioritised, lack of strategy approach in the planning of metallurgy development and neglecting the priorities of post-industrial society, principles of sustainable development and other life priorities by governing bodies, against a background of economic stagnation and the never-ending struggle for the redistribution of metallurgical concerns among disputing groups of large owners: all these factors have led to a situation where practically all metallurgical enterprises work mainly for the aims of exporting cast iron and steel at dumping prices all over the world, leaving mountains and torrents of waste in the Ukraine. The proceeds from sales are by no means used for the development or introduction of cleaner or environment-friendly technologies, replacement of out-of-date equipment or the development of long-term programmes aimed at providing environmental safety, social well-being and industrial re-orientation in regions where heavy industry is prevalent.

Radical improvements in the environmental situation and achievement of the aims of sustainable development will only be possible in the Ukraine after a fundamental revision of the government's economic policy in favour of the long-term ecological priorities of the nation and the establishment of reliable public control in this sphere.

225

Acknowledgements

The authors would like to express their kind gratitude to Dr. Temir Britajev, biologist, 'Sjevjercov' Institute of Evolutionary Morphology & Ecology of Animals, Russ. Acad. Sci., Moscow, Russia, Valjerij I. Rodinov, biologist, The Institute of Biology, Acad. Sci., Latvia, Larisa A. Smirnova, ecologist, RosGiProVodHoz, Moscow, Russia, and all their colleagues, who took part in the joint research, and special thanks to Sergij V. Kovalenko, designer, Kharkov.

References

Abramovich, G.V. (1977) Zashchita vodnogo bassjejna pri pjerjerabotkje mjetallurgichjeskih shlakov [Protection of Water Basin under Processing of Metallurgical Slag]. In: *Trudy UralNIIChjerMjet*, vol.29. - Svjerdlovsk, 1977. pp 32–35. *(in Russian)*

Andon'jev, S.V. & Filip'jev, O.V. (1979) *Pyljegazovyje vybrosy prjedprijatij chjornoj mjetallurgiji* [Dust and Gas Emissions from Iron Works], 2nd ed. - Moscow, 'Mjetallurgija' Publ.House, 1979. 192 p. *(in Russian)*

Baryshnikov, V.G., Gorjelov, A.M., Papkov, G.I. et al. (1986) *Vtorichnyje materialnyje rjesursy chjornoj mjetallurgii. V 2-h tomah. Tom 2: Shlaki, shlamy, othody obogashchjenija zhjeljeznyh i margancjevyh rud, othody koksohimichjeskoj promyshljennosti, zhjeljeznyj kuporos (obrazovanije i ispolzovanije): Spravochnik* [Secondary Row Material Resources of Ferrous Metallurgy. In 2 vols. Vol.2: Slag, Slurry, Wastes from Iron Ore and Manganese Ore Dressing, Wastes from By-Product-Cocking Industry, Green Vitriol (Production and Utilization): Reference Book]. - Moscow, 1986. 344 p. *(in Russian)*

Britajev, T. *Personal communication.* July 2002.

Brjeslavjec, A.I., Malyzhjenkova, V.V., Svirenko, L.P. et al. (1985) Ob ispol'zovaniji martenovskih shlakov v drjenazhnyh sooruzhjenijah kanala po pjerjebroskje chasti stoka Volgi v Don [About the Use of Martin Slag in Drainage Constructions of the Canal for Transfer of the Part of Volga River Flow to the Don River]. In: *Shlaki chjornoj mjetallurgiji i sposoby ih pjerjerabotki: Collected articles.* - Svjerdlovsk, VNIIChjerMjet, 1985. pp.101-109. *(in Russian)*

Bronfman, A.M. (1972) Sovrjemjennyj gidrologo-gidrohimichjeskij rjezhim Azovskogo morja i vozmozhnyje jego izmjenjenija [Contemporary Hydrological and Chemical Regime of the Sea of Azov and its Possible Changes]. In: Rybohozjajstvjennyje issljedovanija Azovskogo morja: *Trudy AzNIIRH, vyp.10.* - Rostov-na-Donu, 1972. pp 20-40. *(in Russian)*

Bronfman, A.M. & Hljebnikov, Je.P. (1985) *Azovskoje morje: Osnovy rekonstrukcii* [The Sea of Azov: Baselines for Reconstruction]. - Ljeningrad, GidroMjetjeoIzdat, 1985. 265 p. *(in Russian)*

Bryginjec, Je.D., Saratov, I.Je., Svirenko, L.P. & Rjemizov, V.I. (1999) Zashchita Azovskogo morja ot zagrjaznjenija othodami kombinata AzovStal [Protection of the Sea of Azov from Pollution by Wastes from AzovStal Combine Works]. In: *Visnyk Ukrajins'kogo Budynku ekonomichnyh ta naukovo-tehnichnyh znan' tovarystva "Znannja" Ukrajiny*, № 3, 1999. P.137-142. *(in Russian)*

Chudakov, F.Ja. & Tihonova, N.N. (1985) K mjehanizmu poroobrazovanija v mjedljennoostyvajushchjem domjennom shlakje [A Contribution to the Mechanism of Pore

Forming in Slow-cooling Blast Furnace Slag]. In: Svojstva shlakov chjornoj mjetallurgiji i sposoby jih pjerjerabotki: *Trudy VNIIChjerMjet.* - Svjerdlovsk, 1985. P. 33-38. *(in Russian)*

Fix, W., Moradoghli-Haftmani, A. & Schwerdtfeger, K. (1975) Über die Porigkeit von Hochofenschlacken unter Beruchsichtigung des Eintlusses von gelosten Stickstoff. In: *Arch. Eisenhuttenwes,* v. 46, №6, 1975. pp 363-370.

Futamura, T. (1981) Mechanism of the Porous Blast-furnace Slag Formation and the Dense Slag Production. In: *Nippon Steel Tech. Rep.,* №7, 1981, pp.9-17.

Hejný, S. & Sytnik, K.M.(eds.) (1993) *Makrofity - indikatory izmjenjenij prirodnoj srjedy* [Macrophytes as Indicators of Environmental Shifts]. - Kijev, 'Naukova dumka' Publ. House, 1993. 434 p. *(in Russian)*

Hrustaljov, Ju.P. et al.(1981) Zakonomjernosti rasprjedjeljenija i dinamika marganca, vanadija, mjedi i nikjelja v vodnoj tolshchje Azovskogo morja [Patterns of Distribution and Dynamics of Manganese, Vanadium, Copper and Nickel in Water Column of the Sea of Azov]. In: *Gjeografichjeskije aspjekty izuchjenija gidrologiji i gidrohimiji Azovskogo morja: Collected articles.* - Ljeningrad, Gjeografichjeskoje obshchjestvo SSSR, 1981, pp58-66. *(in Russian)*

Jakubovskij, K.B., Mjerjezhko, A.I. & Njestjerjenko, N.P. (1975) Nakopljenije vysshymi vodnymi rastjenijami eljemjentov minjeralnogo pitanija [Accumulation of Nutritive Elements by Higher Aquatic Vegetation]. In: *Biologichjeskoje samoochishchjenije i formirovanije kachjestva vody.* - Moskva, 'Nauka' Publ. House, 1975, pp 17-62. *(in Russian)*

Karjakin, L.I. (1949) Mjetallurgichjeskije shlaki v sorjemjennyh morskih otlozhjenijah [Metallurgic Slags in Contemporary Marine Sediments].In: *Priroda,* №6, 1949. P. 26-28. *(in Russian)*

Kisiljevskij,V.V., Bocharov, V.A. & Rossoshanskaja, V.L. (1980) Oprjedjeljenije tjazhjolyh mjetallov v stochnyh vodah i osadkah [Identification of Heavy Metals in Waste Waters and Sediments]. In: *Trudy konfjerjencii "Uskorjennyje mjetody himichjeskogo kontrolja v promyshljennosti".* - Donjeck, 1980, pp. 177-178. *(in Russian)*

Kormyshjev, V.V. et al. (1981) Issljedovanije haraktjeristiki parogazovyh vybrosov pri poluchjeniji gravijepodobnoj pjemzy iz domjennyh shlakov [Investigation of characteristics of steam and gas emissions when producing gravel-like pumice-stone of blast furnace slag]. In: *Pjerjerabotka i ispol'zovanije domjennyh, staljeplavil'nyh i fjerrosplavnyh shlakov: Trudy UralNIIChjerMjet.* - Svjerdlovsk, 1981, pp. 42-47. *(in Russian)*

Lapchinskaja, L.V., Smyslova, L.I. et al. (1988) Emissionnaja spjektroskopija kak mjetod izuchjenija komponjentov biosfjery [Emission Spectroscopy as a Method for Investigation of Biosphere Components]. In: *Vjestnik Khar'kovskogo Univjersitjeta,* № 325. - Kharkiv, 'Vyshcha Shkola' Publ.House,1988, pp. 60-65. *(in Russian)*

Liepa, R., Rodinov, V., Cimdiņš, E. et al. (1989) The Present Ecological State of the Lake Liepaja. In: *Latvijas PSR Zinatnu Academijas Vestis,* №5, 1989, pp. 70-78.

Litvinova, T.I., Pirozhkova, V.P. & Pjetrov, A.K. (1972) *Pjetrografija njemjetallichjeskih vkljuchjenij* [Petrography of Non-metallic Inclusions]. - Moskva, 'Mjetallurgija' Publ. House, 1972, p.157 *(in Russian)*

Mjetodichjeskoje rukovodstvo po analizu tjehnologichjeskih i stochnyh vod chjornoj mjetallurgiji [The Methods Manual on Processing and Waste Waters Analyses in Ferrous Metallurgy]. - Moscow, 'Mjetallurgija' Publ. House, 1988. 359 p. *(in Russian)*

Olpinski, K.& Christensen, C. J. (1981) Slope Protection along St. Lawrence Seaway Canals. In: *Can. Geotechnical Journ.*, v.18, №3, 1981, pp. 402 – 419.

Pjerjepjelicyn, V.A. (1987) *Osnovy tehnichjeskoj minjeralogii i pjetrografii* [Fundamentals of Technological Mineralogy and Petrography]. - Moscow, 'Njedra' Publ. House, 1987, p. 255 *(in Russian)*

Sajdullin, R.A. et al. (1979) Issljedovanije vlijanija nitrida aljuminija na porizaciju shlakovyh rasplavov [Investigation of the Effect of Aluminium Nitride on Pore-forming in Slag Melt]. In: *Shlaki chjornoj metallurgiji: Trudy UralNIIChjerMjet, v.35.* - Svjerdlovsk, 1979, pp.63-67. *(in Russian)*

Sehi, A., Aso, J., Okubo, M. et al. (1986) Development of Dusting Prevention Stabilizer for Stainless Steel Slag. In: *Kawasaki Steel Tech. Rep.*, №15, 1986, pp.16-21.

Shkolnik, Ja.Sh. et al. (1985) Sokrashchjenije vybrosov sjernistyh gazov pri granuljaciji domjennyh shlakov [Reduction in Emissions of Sulfur Oxides when Granulating Blast Furnace Slag]. In: *'Stal'*, №1, 1985, pp.83-95. *(in Russian)*

SNiP: Rukovodstvo po projektirovaniju sanitarno-zashchitnyh zon promyshljennyh prjedprijatij [Building Standards and Regulations - Guidelines on Design of Sanitary Protection Zones for Industrial Enterprises]. - Moscow, StrojIzdat Publ. House, 1984. 78 p. *(in Russian)*

Shatrovskij, A.G. & Spirin, A.I. (2002) Jestjestvjennaja ijerarhija urbanistichjeskih sistjem [A Natural Hierarchy of Urbanistic Systems]. *In: 'Kommunalnoje hozjajstvo gorodov'.* Vyp. 36. Sjerija "Arhitjektura i tjehnichjeskije nauki". - Kijev: "Tehnika" Publ. House, 2002, pp. 173-177. *(in Russian with English abstract)*

Stankovych, V.V. (sci. head manager) (1983) *Gigijenichjeskaja ocenka vlijanija utilizirujemyh othodov domjennogo proizvodstva na sanitarnyj rjezhim r. Dnjepr v rajonje g. Dnjeprodzjerzhinska: Otchjot o NIR (zakljuchitjel'nyj)* [Hygienic Assessment of Utilised Blast Furnace Wastes on Sanitary Regime of the Dnieper River off the city of Dnjeprodzjerzhinsk: Research Report (final)]. Theme 07.01.0014.305.83. Kijev Research Institute of General and Municipal Hygiene. - Kijev, 1983, p. 33 *(in Russian)*

Stolberg, F.V., Ladyzhjenskij, V.N. & Spirin, A.I. Bioplato – (in press) effjektivnaja malozatratnaja ekotjehnologija ochistki stochnyh vod [Constructed Wetlands as Efficient and Cost-effective Ecotechnology for Wastewater Treatment]. *Proc. Intern. Conference "Interregional Problems of Environmental Safety, IPES'2002"*, 15-16 May 2002, Sumy, Ukraine. - Sumy, 2002 (in press) *(in Russian with English abstract)*

Svirenko, L.P. (1976) Svojstva metallurgichjeskih shlakov Ukrajiny i ih ispol'zovanije v gidrotjehnichjeskom stroitjel'stvje [Properties of Metallurgical Slags from Ukraine and their Use in Hydraulic Works Construction] In: *Mjetody issljedovanija iskusstvjennyh gruntov dlja stroitjel'stva: Collected articles.* – Kyjiv, 'Znannja' Publ. House, pp. 97-105.*(in Russian)*

228

Svirenko, L.P. (1978) Shlakoobrazujushchije minjeraly i gidravlichjeskaja aktivnost' staljeplavil'nyh shlakov [Slag-forming Minerals and Hydralic Activity of Steel Slags] In: *Trudy UralNIIChjerMjet, Vyp. 32.* – Svjerdlovsk, 1978, pp. 111-115. *(in Russian)*

Svirenko, L.P., Lapchinskaja, L.V. & Sivtsova, A.V. (1990) Vlijanije ugljedobychi i othodov metallurgichjeskogo proizvodstva na gidrohimiju rjek i vodohranilishch [Impact of Coal Mining and Metallurgical Wastes on Water Chemistry of Rivers and Reservoirs]. In: *Vjestnik Khar'kovskogo Univjersitjeta, № 345.* – Kharkiv, 'Vyshcha Shkola' Publ.House, 1990, pp. 78-83. *(in Russian)*

Svirenko, L.P. & Saratov, I.Je. (1983) Naturnyje issljedovanija damby iz shlaka v morje [Field Investigation of Slag-formed Dam in the Sea] In: *Trudy VodGeo.* – Moskva, 1983, pp. 154-161. *(in Russian)*

Svirenko, L.P., Saratov, I.E. & Lapchinskaja, L.V. (1991) Vlijanije sbrosa othodov mjetallurgichjeskogo proizvodstva na vodnuju ekosistjemu oz. Liepajas, Latvija [The Effect of Dumping of Wastes of Metallurgical Industry on Aquatic Ecosystem of the Liepajas Lake, Latvia] In: *Vsjesojuznaja confjerjencija "Gydromehanizirovannyje raboty i damby"*: Tezisy dokladov. – Rostov-na-Donu, 1991, pp. 237-238. *(in Russian)*

Svirenko, L.P., Saratov, I.Je. & Rjemjezov, V.I. (1992) *Avtorskoje Svidjetjel'stvo* [Inventor's Certificate] SU 1781368 SSSR, MKI E 02 B 7/06. 1992 *(in Russian)*

Svirenko, L.P. & Spirin, A.I. (1995) Prevention of Eutrophication and Pollution of Aquatic Ecosystems when Using and Dumping Slags. In: *7th European Ecological Congress "Ecological Processes: Current Status and Perspectives"./ Book of Abstracts* /A. Demeter & L. Peregovits (eds.) - Budapest, Hungarian Biological Soc., 1995, p. 137.

Svirenko, L.P. & Spirin, A.I.. (1997) Wetlands of Ukraine: the National Economy vs. the Environment. In: *Conflict and the Environment* / Eds. N. P. Gleditsch et al. - Dordrecht, Kluwer Acad. Publ., 1997, pp. 451-470. (NATO ARW Series)

Svirenko, L.P. & Yurchenko, V.A. (1995) Biotic Impact on Metal Distribution in Brackish Waters from Metallurgical Works. In: *7th European Ecological Congress "Ecological Processes: Current Status and Perspectives"./* Book of Abstracts /A. Demeter & L. Peregovits, (eds.) - Budapest, Hungarian Biological Soc., 1995, pp. 132.

Tarabrin, V.P., Chjernyshova, L.V. & Pjel'tihina, R.I. (1984) Ispolzovanije zjeljonyh nasazhdjenij dlja optimizaciji srjedy v zonje zagrjaznjenij prjedprijatij chjornoj mjetallurgiji [Use of Green Stands for Environmental Optimisation in the Zone of Pollution Caused by Iron Works]. In: *"Rastjenija i promyshljennaja srjeda": Sbornik statjej.* - Svjerdlovsk, Ural'skij gosudarstvjennyj univjersitjet, 1984, pp. 101-106. *(in Russian)*

Verma, B. P. (1982) Suitability of Slag for Construction of Embankments and Dams In: *Proc. of the Conference on Construction Practices and Instrumentation in Geotechnical Engineering (Surat, India, December 20-23, 1982).* - Surat (Gudjarat), 1982, pp. 61-67.

Vinogradov, M. E. (ed.) (1991) *Izmjenchivost' ekosistjemy Chjornogo morja: jestjestvjennyje i antropogjennyje faktory* [Changeability of the Black Sea Ecosystem: Natural and Man-made Factors]. - Moskva, 'Nauka' Publ. House, 1991, p.427 *(in Russian)*

Zarubov, A., Parele, E. & Liepa, R. (1989) Effect of Hydrochemical Differentiation on the Composition and Structure of Zooplankton and Benthos in the Lake Liepaja. In: *Latvijas PSR Zinatnu Academijas Vestis*, №5, 1989, pp. 79-86.

Zolotuhin, I.A., Nikulina, S.N. & Fjedosjejeva, L.A.(1995) Snizhjenije koncjentracii mikroeljemjentov v vodnoj srjedje pod vozdjejstvijem kornjevyh sistjem [Decreasing Concentrations of Microelements in Aquatic Environment under Effect of Root Systems]. In: *Ekologija (Ecology)*, №3, 1995, pp. 248-249.

CHAPTER 23
Perspectives in utilizing ash and slug wastes from the chemical industry, metallurgy and power generation for environmental purposes

Prof. Mikhail Ginzburg[1], Arnaud Delebarre[2], Alan G. Howard[3], Ilyushchenko Mikhail[1], Lebedeva Olga[1], Mazhrenova Nailya[1], Sarmurzina Alma[1]

[1] Institute of New Chemistry Technologies and Materials of Al Farabi Kazakh National State University, 95a Karasay Batyr Str., Almaty 380012, Kazakhstan

[2] Ecole des Mines de Nantes, Department of Energy and Environmental Engineering, La Chantrerie 4, rue Alfred Kastler, BP20722, F-44307 Nantes Cedex 3, France

. [3] University of Southampton, Department of Chemistry, Southampton, SO17 1BJ, United Kingdom

Summary

Ash slag, formed as a result of industrial processes in the metallurgical, chemical and power generation industries, is only used in significant amounts in the production of cheap construction materials and road covering. At the same time, slag is the potential source of silicate raw material not only for the production of construction materials but also for sorbents and catalysts. The experiments were carried out with two types of ash with different chemical compositions: silico-aluminum fly ash and calcic fly ashes, and also with electrothermo-phosphoric and metallurgical slag. Both sorption and desorption of mercury and copper salts, recovery of valuable components from fly ash and methods of creation of an efficient sorption surface were studied. This paper demonstrates the principle means of using slag porous materials as substrates of catalysts.

Introduction

Slag dumps from metallurgical and chemical production and power generation are a major ecological problem in Kazakhstan. Transportation of slag to dumps and maintenance of the dumps uses up significant resources.

At present, the main application of industrial slag is in displacing original mineral raw materials in production of materials for construction work and as catalysts (V.G Panteleev et al., 1978). Attitudes towards the use of slag have changed in recent years: waste materials are secondary material resources and, as such, valuable raw material for other industries. Slag is a potential source of silicate raw material, not only for the production of construction materials but also for sorbents and catalysts as well.

Metallurgical blast-furnace slag and the materials which have this as their basis are good sorbents for cleaning polluted industrial wastewater (T.I.Priimak et al., 1982).

The principle area in which slag silicate sorbents are applied is in the treatment of multicomponent industrial wastewater where it is used for the removal of cations of nonferrous metals. Slag utilization is complicated because phase structure varies, as does the size and distribution of

231

W. Leal Filho and I. Butorina (eds.),
Approaches to Handling Environmental Problems in the Mining and Metallurgical Regions, 231–245.
© 2003 *Kluwer Academic Publishers. Printed in the Netherlands.*

particles, and the chemical composition, and also because there is a lack of industrial methods of processing it. Hence, problems arise because it is difficult to obtain materials with constant and predictable characteristics arise regards their physico-chemical properties and metal ions adsorption at the liquid/solid interface.

The processes of metal ion adsorption at the liquid/solid interface as well as the application of so-called low-cost sorbents have been an object of research for several years.

The aim of the present research is to study a means of using waste products from the metal, chemical and power generation industries as sorbents for industrial wastewater treatment from metals, particularly from mercury, as substrates of catalysts for major industrial processes, and for recovering valuable components from wastes. Our major focus was on the adsorption of mercury.

Mercury is known for its toxicity towards the aquatic environment. The discharge of effluents containing mercury in the environment can constitute a threat to living organisms in an aquatic environment and have serious repercussions on the food chain. The extreme toxicity, the bio-accumulation of this element in soils and sediments as well as numerous major environmental accidents, especially between 1953 and 1956 on the shores of Minamata Bay in Japan have been instrumental in the development of treatment processes of effluent containing mercury (E.P.Yanin, 1997)

Kazakhstani fly ash is a good sorbent for heavy metals, especially for mercury, even on a commercial scale. For example, the industrial release of mercury from the Synthetic Rubber Factory in Temirtau, Kazakhstan into the Nura River was adsorbed by ash from the power plant in Temirtau (S.Heven et al., 2000).

Initially, the adsorption by fly ash of mercury present in aqueous solutions was studied in a static-mode reactor, and the catalyst potential was measured simultaneously. Following this, a test was carried out to estimate the capacity of solids to retain durably mercury ions   =?to retain mercury ions over a long period of time. The research has also shown that electrothermophosphoric slag and ferroalloy slag can be used as sorbents of metals and substrates for catalysts after special processing. The possibility of using fly ash as raw material for obtaining valuable components was evaluated. The final step, after completion of the adsorption experiment, was to investigate the surface of spent adsorbents, in order to obtain an understanding of the mechanisms involved in mercury adsorption.

Materials and Methods

Characteristics of original materials.

The research was conducted on wastes from the electrothermal production of phosphorus - phosphoric slag, fly ashes from Ekibastuz and Krasnoyarsk and ferroalloy slag of AO "Ferrohim".
The chemical composition of the materials is given in Table 1.

Table 1 Chemical composition of wastes (weight %).

Wastes	Chemical composition, %						
	SiO$_2$	Al$_2$O$_3$	Fe$_2$O$_3$	MgO	CaO	P$_2$O$_5$	Cr$_2$O$_3$
Phosphorus slag	37.0-40.0	2.0-2.5	0.3-0.4	5.0-10.0	40.0-42.0	6.0-10.0	
Ferroalloy slag *	28.0-30.0	4.0-8.0	1.0 – 2.0	6.0-40.0	48.0-55.0		4.0-7.0
Fly ash from Ekibastuz coal	45.0-53.0	21.0-24.0	3.0-12.0	1.0-2.0	2.0-10.0		
Fly ash from Krasnoyarsk coal	34.0-38.0	8.0-10.0	11.0-12.0	7.0-8.0	32.0-35.0		
Silico-aluminum fly ash (France)	44.9	22.0	8.4	3.5	5.7	0.1	
Sulfo-calcic fly ash (France)	43.5	18.0	7.4	1.4	16.9	0.4	

* The data was found in the paper V.Grinenko, B. Nurtaeva et al. Utilization of waste products of Actyubinsk Ferroalloy Plant. Promyshlennost Kazakhstana, 2000, №12, P.94-95

Two types of fly ash with different chemical compositions were tested in this study: silico-aluminum and calcic fly ashes, as was electrothermophosphorus and metallurgical slag. Fly ash from France which was similar in chemical composition, was used to study mercury desorption.

The Experiment

A series of experiments was carried out: an Hg-adsorption kinetics series with simultaneous measurement of potential and an Hg-desorption isotherms series. The aim of the experiments was to determine the mechanism of mercury adsorption by fly ash. The initial mercury ion concentration was chosen as 1×10^{-5}- 5×10^{-5} mol/l (20-100 mg/l), and the ash concentration was 25 g/l. Solutions and ash were shaken at 250 rpm in a thermocontroled bath at 25°C or 30°C (desorption). During the experiment pH levels of solutions were adjusted to three set values 2.5, 5.0 and 7.5 or 3, 4 and 5 (desorption) by addition of an acid. The samples were filtered through 0.45-μm pore size membrane filters. The experiments were then carried out to develop methods of obtaining porous slag materials by acid-base and mechanical activation and of applying them as substrates of catalysts and sorbents.

Experiments on recovery of valuable components from fly ashes were also carried out.

Analytical methods

The analysis of mercuric and copper ions in solution was carried out using Perkin-Elmer atomic adsorption spectrometry (Perkin-Elmer, 1997). Macro component composition of slag and slag porous materials was determined by chemical analysis using standard methods. The standard error for the analyses of the majority of elements did not exceed 5 %, for determination of phosphorus the standard error was 10%. Phase composition of samples was studied by roentgen-structural analysis using a DRON-5 instrument with Cu-C α-radiation.

The specific surface area of samples during routine analyses was determined by express method of nitrogen thermo-desorption (in reference mode). The relative error, estimated on 5-7 replicates, did not exceed 10 %. For a number of samples more exact measurements were made using the BET method; the calculations were carried out on nitrogen desorption branch of low temperature isotherm.

Pore size distribution was estimated on a "Porosimeter 2000/ Carla Erba" by the mercury impression technique, and also by low-temperature adsorption of nitrogen.

Registration of infrared spectra was carried out using a Specord-75IR.

Particle sizes were determined by transmission electron microscopy.

Mechanic-chemical processing was performed in a high-energy spherical mill, which is a vessel made of stainless steel without heat removal. The rate of the vessel's alternate motion depended on the load, the average rate being 1000 oscillations per minute. The intensity of processing varied by change of loading of the mill with steel spheres =? when the mill was loaded with steel spheres.

Results and Discussion

a) Mercury adsorptions by fly ash

Fly ash is a product of fuel burning. This system has a complex chemical and phase composition, including oxides of aluminum, silicon, iron, titanium, potassium, sodium, calcium, magnesium, and organic substances. All these components are products of both thermochemical and phase transformations of organic and mineral constituents of coal. Fly ash composition includes carbon, which has unique affinity to Hg (II) complexes (G.S.Polimbetov, Ya.A.Dorfman, 1998) and forms strong links with them. This fact makes the study of mercury adsorption by fly ash more complicated.

Mercury adsorption by fly ashes is also complicated by the concurrent process of mercury desorption and the formation of relatively insoluble compounds of mercury. Figure 1 demonstrates the extent of $Hg(NO_3)_2$ adsorption and its dependence on the equilibrium concentration of fly ashes from Ekibastuz coal. The higher the fly ash equilibrium concentration, the lower the rate of $Hg(NO_3)_2$ adsorption. It can be seen that equilibrium of adsorption reaction was almost reached after 5 hours, for both types of ash. The rate of adsorption was as high as 40 %. The rate of adsorption for calcium fly ash is higher than for silico-aluminum at all values of pH in aqueous solution.

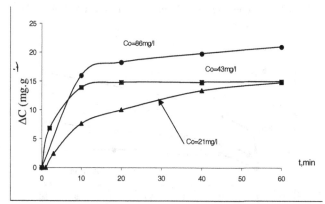

Fig. 1. Dependence of mercury adsorption rate on initial mercury concentration on silico-
aluminous ashes. $[Hg^{2+}]o$ varied from 21mg/l to 86 mg/l, m = 0.5 g, t = 25°C.

Adsorption corresponds to Freundlich isotherm, ranging in concentration from 1.3 x 10^{-3} to 2.6 x
10^{-3} mol/l.

The shape of adsorption isotherms for both types of ash is identical, which indicates that identical
mechanisms could be responsible for mercury adsorption. Apparently, different mechanisms of
adsorption occur at different concentration. This is also proved by potentiometric curves (Figure 2).

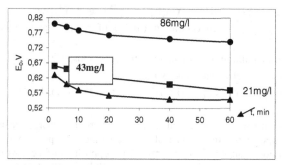

Fig. 2. Potentiometric curves of dependence of mercury adsorption on initial mercury concentration
in silico- aluminous ashes. $[Hg^{2+}]o$ varied from 21mg/l to 86 mg/l, m = 0.5 g, t = 25°C

In silico-aluminum ashes (Ekibaqstuz) the potential shifts to anode region and reaches 50 mV. The
lower are initial concentration and duration of the process the higher is potential shift to anode
region. On the contrary the potential shifts symbately to the initial concentration for calcium ashes
(Krasnoyarsk) and does not practically depend on duration of the process.

The pH level of the solution has a considerable influence on the adsorption process. When ash is
added, the solution becomes more alkaline. With an increase in pH, the surface charge, the
composition of mercury compounds in the solution, and the stages of mercury adsorption by the ash
are altered (M. Bek, I. Nadypal, 1989).

236

This can be clearly seen in potentiometric (Figure 2) and kinetic (Figure 3) curves of mercury adsorption.

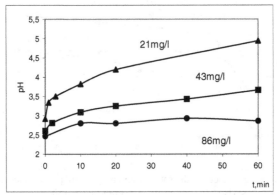

Fig. 3. Dependence of pH change on initial mercury concentration on silico- aluminous ashes. $[Hg^{2+}]o$ varied from 21mg/l to 86 mg/l, m = 0.5 g, t = 25°C.

The addition of ash to an aqueous solution of Hg (II) salt results in the potential shift to 0,02-0,43V, depending on the nature of ash.

According to the Nernst equation, the redox potential (φ) for general semi-reaction
$$2Hg^{2+} + 2e + aA \leftrightarrow Hg_2^{2+} + bB,$$ measured in reference to a standard hydrogen electrode, is:
$$\varphi_{ox/red} = 0.911 - RT/2F \ln YHg_2^{2+} [Hg_2^{2+}]Y^b_B[B]^b / Y^2_{Hg}^{2+} [Hg^{2+}]2Y^a_A[A]^a$$
where $[Hg_2^{2+}]$ and $[Hg^{2+}]$ - concentration (mol/l) of reduced and oxidized mercury forms respectively;
[B] [A] – the same for auxiliaries (hydrogen ions, hydroxyl ions, water molecules etc.);
Y – correspondent molar coefficients of activity;
0,911 – standard redox potential of pair Hg^{2+}/Hg_2^{2+} (Ya.A. Dorfman, 1984).

Chemico-mineralogical composition of ash (figure 4) determines charge localization on a surface. In small surface charges, when the number of ionized surface groups ≡MeOH is insignificant, two charges of mercury ions can not be compensated by opposite surface charges because of steric obstacles. For this reason, the oppositely charged ions will be involved into a double layer to retain electrical neutrality (E.A.Nechaev, 1989). Adsorption of mercury cations is accompanied by simultaneous absorption of anions in the absence of specific interactions. Mercury ions adsorbed on the negatively charged centres displace hydrogen ions from the double electrical layer into a solution. During the adsorption of mercury ions there is probably no complete compensation of their charges by the surface charged groups. Adsorption of one ion of mercury is accompanied by one ion of hydrogen arising in a solution. When the pH increases, the number of charged groups grows and a condition is created where some of the double-charged ions can be adsorbed with saturation of both charges, i.e. specific adsorption occurs by creating chemical links between cations and charged centres of the surface.

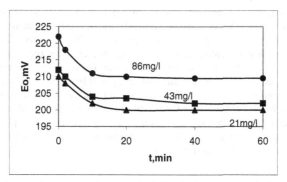

Fig. 4. Potentiometric curves of dependence of mercury adsorption on initial mercury concentration on calcic ashes. [Hg^{2+}]o varied from 21mg/l to 86 mg/l, m = 0.5 g, t = 25°C.

It would appear that adsorption of the mercury cations occurs as follows:

Adsorption of the cations from strong alkaline solutions takes place on the charged surface centres. Thus, a double-charge of mercury cation is compensated by negative charges of the surface. There is specific adsorption with formation of a surface chemical compound. When the pH becomes acid, the proportion of cations forming the link with only one charged group, increases. The second charge is compensated by the charge of an electrolyte. It is quite possible that, when there are further shifts in the pH value, there arises a kind of mercury cation adsorption, which is not connected to the presence of a double electric layer. It seems that, where there is a low pH, mercury ions are adsorbed as neutral complexes such as [Hg(Cl)$_2$], where both neutral atoms and the group of surface atoms log in the coordination sphere simultaneously (Anderson, 1973). Mercury concentration is a very important factor (diagram of mercury species formation, Figure 5). The formation of complexes is favoured by the presence of [Hg(Cl$_4$)]$^{2-}$ and [Hg(Cl$_3$)]$^-$ in adsorption processes (Bonnisel-Gissinger et al., 1999) and it is completed with epitaxial crystallization, i.e. with the new destructive-epitaxial mechanism of adsorption. Difference between solubility of substances participating in the reaction is the driving force of the process in the given adsorption mechanism (Alexovsky, 1990).

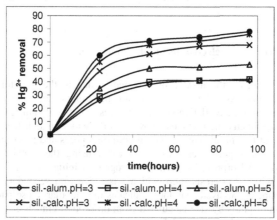

Fig. 5. Kinetic curves of mercury removal onto silico-aluminous and sulfo- calcic fly ashes. [Hg^{2+}]o = 602 mg/l; m =100 g/l; t = 25°C; V = 500 ml.

The results of A. Delebarre's research into the process of mercury adsorption by ash using X-photon spectrophotometry (XPS) analysis proved that oxides of silicon and aluminum are responsible for mercury adsorption on the surface of ash. The mercury adsorption mechanisms correspond to its chemical reactions with various oxides present on the surface of ash. Hydrated calcium silicate $Ca_3Si_3O_9$ is present in adsorption material. However, this hydrate has not resulted in increased adsorption capacity of the fly ash and has, possibly, increased the stability of adsorbed mercury by the formation of a layer of hydrated calcium silicate on the surface. Silicon hydration, the presence of aluminum oxide and the formation of the layer of hydrated calcium silicate are the phenomena that occur during mercury adsorption. As a result passing of pozzolanic reactions in adsorptive material and covering the surface with hydrated calcium silicate can be responsible for the resulting stability of the adsorption material.

The experiments carried out showed that it is possible to use fly ash as a cheap and efficient adsorbent for mercury. Further studies aimed to investigate the possible formation of methylmercury and to develop the technology of stabilization of ash containing heavy metals, particularly mercury (Figure 6).

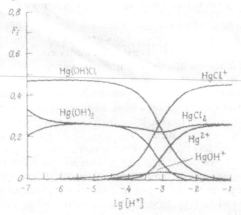

Fig. 6. Diagrams of mercury specious formation in system Hg^{2+}-Cl^- - OH^- (from Beck M. and I. Nagypal, 1989.)

b) Porous slag materials from electrothermophosphoric slag

There is hardly any data on the texture of phosphoric slag in the literature. The study of the characteristics of samples has led to the conclusion that initial slag belongs to the low-porous materials (SSA = 10 ± 2 m^2/g, cumulative volume of pores – 0.16 ± 0.02 cm^3/g). Measurements taken using the mercury porosimetry method have shown that the wide-range distribution of pore size is a characteristic feature of the initial slag, including both mesopores and macropores with a radius of 1000-10000 nm. The mesopore area was investigated using the method of low-temperature adsorption of nitrogen. Adsorption isotherm of nitrogen on initial slag has a similar pattern for mesopore materials with a hysteresis loop, which arises due to a secondary process of capillary condensation; the shape of the hysteresis loop indicates the globular, close-packed structure of the slag. The distribution of pores by size calculated from desorption branch of isotherm has a narrow maximum with diameters in the range of 4-5 nm.

The acid treatment results in a repeated increase in the specific area of phosphoric slag. The comparison of poregrams of initial and modified slag leads to the conclusion that during acid treatment there is an increase in the cumulative volume of pores and a gradual redistribution of pores by size: the proportion of large pores with a radius greater than 1000 nm decreases, and the proportion of mesopores increases (Table 2).

Table 2 Texture characteristics of samples obtained by 1M H_3PO_4 treatment of slag after certain time intervals

Treatment time, hours	Specific surface area, m^2/g	Cumulative volume of pores, cm^3/g
1	29	0.17
2	69	0.21
3	116	0.39

Of particular interest from our point of view is the fact that micropores are detected in slag samples treated by acids. The average diameter of micropores lies in the range of 6-8 Å, their cumulative volume is about 0,03 cm^3/g.

The chemical basis of the mechanism of acid pickling of slag glasses is the ion-exchange replacement of alkaline-earth metals by protons. This is confirmed by a large volume of the data. So the treatment of slag by acids results in changes in its infrared-spectrum in the range of 600 - 400 cm^{-1} , the changes becoming more significant with increased treatment time. The absorption in this range is caused by vibrations of metal-oxygen polyhedrons $(CaO)_n$, and also by deformation vibrations of bridge and end groups. The changes in infrared-spectra can be related to the gradual removal of calcium ions.

The data resulting from the classical chemical analysis of the samples also shows removal of calcium and magnesium cations during acid treatment.

The ability of slag's calcium cations to ion exchange is also confirmed by the results of experiments involving treatment of slag with solutions of ammonium and sodium salts.

It appears that the chemical transformations in slag materials under acid treatment are not only limited by the replacement of alkaline-earth cations by protons. There is a further possibility of the condensation of closely located groups $\equiv$ Si-OH and the change the structure of silico-oxygen lattice caused by that. However, even this fact does not explain the development of a mesopores structure.

The most probable explanation is heterogeneity and two-phase nature of slag glasses. The treatment of such glasses by acids is accompanied by destruction of the entire structure occupied by the phase less resistant to acids, thus creating conditions for mesopore formation.

The chemical composition and the properties of porous slag materials are altered in a wide range depending on the extent of acid treatment of the slag. For the directional adjustment of properties it was necessary to establish trends in the way they changed during acid treatment.

Acid modifying of slag and the influence of various factors (i.e. acid nature, concentration, time of treatment and temperature) on this process was studied in detail. Three acids - hydrochloric, nitric and phosphoric were chosen for the study. It should be noted that these acids are divided into two different types according to the nature of their influence on classical silicate glass: hydrochloric and nitric acids are considered to be instrumental in the removal of the silicate hydrolyze and alkaline component the solution, whereas phosphoric acid destroys the silico-oxygen lattice of the glasses layer by layer.

Table 3 demonstrates the influence of the nature of the acid on the chemical composition of slag material. The effect of 1.5M acid solutions is used as an example. The change in composition occurs mainly due to the decrease in weight percentage of calcium and magnesium oxides, the weight percentage of SiO_2 increases simultaneously. It is interesting to note that the iron content hardly changes and that a significant portion of aluminum remains in the porous materials. Hydrochloric acid clearly has the greatest impact on the studied reagents.

The opportunity to adjust the specific surface area of prepared porous materials is particularly interesting. It has been established that this could be adapted to include a wide range (10-300 m^2/g) by varying the conditions of treatment with acids.

Table 3. Influence of treatment by different acids on macro component composition of slag porous materials (treatment time - 1 hour, concentration of acids - 1,5 equivalent per litre).

Acids	Content, % (weight)					
	SiO_2	CaO	MgO	Fe_2O_3	Al_2O_3	P_2O_5
Initial slag	40.8	41.4	10.1	0.3	2.0	10
H_3PO_4	51.3	26.2	1.5	0.4	1.1	5
HNO_3	55.3	30.3	1.7	0.3	1.6	3
HCl	61.8	11.7	1.4	0.3	1.2	3

Slag porous materials were approved as substrates for metallic, oxide and sulphide catalysts. In all experiments a series of catalysts was prepared on the basis of original slag and slag materials treated by acids to varying degrees. All catalysts samples were prepared by the impregnation technique. The following catalysts were studied: copper-slag catalysts for dehydration of cyclohexanol, catalysts for n-heptane conversion, cobalt-molybdenum and nickel-molybdenum slug catalysts for hydrodesulphurization of oil fractions. The latter are reviewed in detail.

The specifications of slag porous materials used as substrates are shown in Table 4. The catalysts were prepared in order to keep the proportions of active components constant in all samples: Co (Ni) - 3%, MoO_3 - 12% by weight.

Table 4 Texture characteristics of the substrates and the catalysts hydrodesulphurization.

	Catalyst	Specific surface area, m^2/g		Cumulative volume of pores, sm^3/g	
		Carrier	Ready catalyst	Carrier	Ready catalyst
Carrier 1	CMS–1	45	7	0.21	0.21
Carrier 2	CMS–3	169	81	0.67	0.56
Carrier 3	NMS	152	54	0.66	0.51

Hydrodesulphurization of diesel fraction was studied using a pilot catalytic device with circulation of hydrogen-containing gas. The data on catalyst activity are presented in Table 5.

Table 5. Activity of cobalt-molybdenum slag and nickel-molybdenum slag catalysts of hydrodesulphurizatin at different temperatures expressed in the residual content of sulphur in a diesel fraction. Vo = 4 hour^{-1}, sulphur content in raw material - 1 % (weight).

Catalyst	Sulphur, % (weight)			
	330°C	350°C	370°C	390°C
CMS−1	0.55	0.37	0.26	0.16
CMS−3	0.36	0.25	0.11	0.06
NMS	0.39	0.25	0.13	0.11

A nickel-molybdenum catalyst is less active than a cobalt-molybdenum catalyst applied on slag material of the same texture. This corresponds to data in the literature on the ratio of activities of traditional aluminum-nickel-molybdenum and aliminum-cobalt-molibdenum catalysts: as a rule, when the conditions are the same, the hydrodesulphurization activity of the former is slightly less pronounced.

The activity of cobalt-molybdenum sludge catalysts strongly depends on the properties of the slag substrate (Table 5). Catalyst CMS-1, prepared on the base of substrate 1 with a low specific surface area and predominantly macroscopic pores in porous structure shows little activity for hydrodesulphurization at all studied temperatures. Catalyst CMS-3 has almost the same activity as traditional catalysts of hydrodesulphurization and is much more active than CMS-1. A detailed study of the porous structure of substrate 2 (which is the basis for preparation of CMS-3) showed that the structure is bimodal. About half the pores have a radius of less than 200 Å, and at the same time macroscopic pores are seen in the structure. Porous structure remains in the prepared catalyst CMS-3.

Results of RPA showed that CMS-3 is roentgen-amorphous. So, it is possible to state that compounds of molybdenum in the catalyst are in an active disperse state. The results obtained show ways of using slag porous materials as catalysts substrates. Waste from the metal industry could also be used as catalyst substrates (e.g. wastes from the Aktubinsk ferroalloys plant). The waste from the Aktubinsk plant was treated with acid and then 0.5% of palladium was applied. The obtained catalyst showed high activity and stability during hydrogenation of wastes from an oil mill (distilled fatty acids of cotton soap stock into general purpose stearic acid). The pressure of hydrogen was 0.1-0.6 Mpa, sparging of excessive hydrogen – 120 h^{-1}, temperature 225°-240 ^{0}C and this catalyst worked in the mode of continuous synthesis of stearic acid for about 700 hours.

The adjustable porous structure of slag materials presents a wide variety of opportunities for their usage as sorbents. Sorption properties of slag-silicate materials in relation to organic compounds are studied in detail for the model system (adsorption of methylene blue). The isotherms of adsorption are created and it was shown that in most cases they correspond to the Langmuir equation. In addition, in the scope of this work the sorption capacity of slag materials for gaseous hydrogen chloride is estimated. Their sorption properties are studied in the In addition, in the scope of this work the sorption capacity of slag materials for gaseous hydrogen chloride is estimated. case of water solutions in relation to heavy metals ions, sulphide-ions as well as technogenic and naturally dissolved organic compounds (DOC).

The sorption of hydrogen chloride by slag porous materials was studied using a laboratory-scale modelling device at specially selected optimal conditions providing the intake of hydrogen chloride into the system and preventing diffusion complexities (the experimental temperature was 30°C, volume rate of gaseous flow was 1500 h^{-1}). It was shown that initial slag has a limited sorption capacity of hydrogen chloride in these conditions. Slag obtained after deep acid treatment (SSA=243 m^2/g) provides 30% of clean-up from hydrogen chloride when its initial concentration in air is 0.27 g/l.

The most effective sorbent among all those studied is the material obtained by sequential treatment of slag by acid and alkali agents (clean-up effectiveness from HCl is up to 50%). It is established that initial slag and slag porous materials can achieve selective adsorption of heavy metals ions from water solutions both in static and in dynamic mode. The effect of acid treatment of slag on the adsorption capacity of obtained materials for copper ions was studied in detail. It was found that, in the case of static mode adsorption, capacity does not, to any noticeable degree, depend on the nature of preliminary treatment of slag. In dynamic mode specific acid treatment increases the duration of effective sorbent activity (the time before skip). The optimum process is a treatment with 2-3 M acids of short duration. Deeper treatment leads to significant worsening of the absorption capacity of the materials.

Sorption of metal ions by slag porous materials is carried out by the destruction-epitaxial precipitation mechanism (Alexovsky, 1990). The process of sorption is a transformation of one low-soluble compound into the other, more stable compound, which is formed directly on the surface of the initial material. Both surface groups and the soluble part of the sorbent could be involved in and define the equilibrium state in the reactions of sorbed ions precipitation. The dissolving of sorbents and the presence of correlation between equilibrium concentrations of sorbed ions and ions of sorbent are the important characteristics of the epitaxial sorption mechanism. The dissolving of silicates from sorbent after it had been in contact with solution for several days was registered and the concentration of silicates was in antibate correlation with the concentration of copper ions.

Slag porous materials were found to be an effective sorbent for heavy metals ions, which is why they can be recommended for use in the treatment of wastewater from non-ferrous metallurgy plants, electroplating workshops and certain other chemical plants. Results obtained when studying the clean-up process from other contaminants of aqueous systems are also positive. The application of sorbents described above in a clean-up from sulphide-ions (with a concentration of 5 mg/l which is common for industrial wastewater) means the concentration can be reduced to the maximum permissible level or even lower. The assessment of the ability of slug sorbents to sorb dissolved organic carbon (DOC) from natural water showed that, if the content of DOC is moderate, then slag-silicate sorbents used in the dynamic mode make it possible to reduce the content of DOC and improve the water quality (reduce the colouration and completely, or partly, remove the odour). The rate of oxidation by permanganate is reduced in1.5-2 times.

Application of slag porous materials could be related to their being heated to high temperatures. Special experiments were carried out to define the composition of gaseous products when heating slag porous materials to 585 ^{0}C in a helium and oxygen atmosphere. Analysis of products was done by mass-spectrometry and the detection limit was $5\cdot10^{-9}$ torr. The only registered gaseous products were water and an insignificant amount of carbon dioxide.

Development of technology for utilization of ash and slag wastes from metallurgical plants

Concentrations and composition of rare and dispersed elements (vanadium, zircon, gallium, rhenium) in slag wastes and ash are comparable with their content in commercial ores. Marco component composition is also quite stable (Table 1). That is why such materials could be regarded as raw materials for the extraction of micro- and marocomponents, in particular, alumina.

We approved several known methods for alumina extraction: acidic, alkali and combined – acidic-alkali.

The most appropriate method in our conditions was an alkali one. The extraction of alumina is performed as follows:

- Transferring of aluminum into the solution as sodium aluminate by heap leaching with 30% NaOH. During this leaching 60-90% of gallium moves into the solution. The separation of aluminum and gallium is difficult due to their similarity. This leads to a considerable loss of gallium.
- Separation of aluminate solution from insoluble residues.
- Formation of insoluble alumosilicate and its separation by filtration.
- Precipitation of aluminum hydroxide.
- Calcination of aluminum hydroxide to alumina at 1200^0 C.

In the first stages it is necessary to pay attention to the preparation of the original material, to weaken the bonds in the material using mechanochemical or radioactive techniques. This is particularly the case for slag from the metal industry.

The solutions formed as a result of leaching varied in terms of the diversity of microelements. Their concentrations were insignificant, so zeolites from Kazakhstan were used for their concentration and further desorption. OUr studies have shown that it is possible to extract 200 kg of alumina, 3-5 kg of non-ferrous metals and 100 g of rare metals from every ton of slag.

Using heap leaching and cheap natural sorbents favours the development of a highly effective technology for extracting a number of components from ash and slag wastes. This technology is feasible and the price of products is low. The slag should be used as a construction material only after extraction of valuable components.

Conclusions

Adsorption of mercury from solutions onto calcic and silico-aluminous ashes was studied. The results have shown that these types of fly ash might behave as an efficient low-cost sorbent for mercury. The principle use of fly ash from old ash lagoons at power stations is shown to be in the remediation of water polluted by mercury.

Fly ash surface analysis showed the existence of silicon and aluminum oxides (silica, mullit and sillimanite), aluminum halogenous and calcium oxide. The XPS analysis showed that adsorption mechanisms of mercury correspond to the chemical reactions with different oxides present on the fly ash surface. Hydration of silica, presence of aluminum oxides and formation of hydrated calcium silicate are the phenomena that occur during mercury adsorption. As a result, reactions in adsorption material pass by the mechanism of destructive-epitaxial precipitation, and surface covering by relatively insoluble compounds can be responsible for the resulting stability of adsorption material.

The research established physico-chemical trends with respect to change in composition, texture and acid-basic properties of slag porous materials caused by a number of factors (i.e. nature and concentration of the acid, time and temperature of treatment). The observed trends allow having influence directly on properties of slag porous materials.

We demonstrated the principle uses of electrothermophosphore slag and slag porous materials as catalyst substrates. Zinc oxide, metal copper and platinum, sulphide cobalt-molybdenum and nickel-molybdenum catalysts are prepared on a slag basis. An efficient cobalt-molybdenum catalyst was obtained using slag porous material as a substrate. The efficiency of hydrodesulphurization of diesel fuel in the presence of this catalyst is more than 90 % at 390°C.

The principle use of slag porous materials as sorbents in removing various types of pollutant (volatile hydrogen chloride, heavy metal cations, sulphide ions, technogenic and natural dissolved organic compounds) from air and water is demonstrated. The trends in the process of heavy metal adsorption by slag porous materials were determined and the destructive-epitaxial mechanism of the process was established.

Furthermore, it was shown that valuable macro- and microcomponents can be recovered from fly ashes and slag.

References

Alexovsky, A.B. (1990) The handbook of permolecular chemistry. Leningrad: University of Leningrad, p. 290

Anderson, J. H. (1973) The local environment of Co (II), Cu (II) and Cr (III) supported on silica gel. J.Catal.,1973, vol.28, №1, pp.76-82

Bek M. and Nadypal, I. (1989) The study of complexes formation by new methods. Moscow: Mir, p. 378

Bonnissel-Gissinger P., Alnot M. et al. (1999) Modelling the adsorption of mercury (II) on (hydr)oxides: II. alpha-FeOOH (geothite) and amorphous silica. Journal of Colloid and Interface Science, 1999, v. 215, № 2, pp. 313-322

Dorfman Ya.A. (1984) Catalysts and mechanisms of hydrogenation and oxidation. Alma-Ata, The Science of KazSSR, p.352

Dorfman Ya.A. and Polimbetova, G.S. (1998) New organic reaction of phosphine in presence of Hg (II) complexes and n-quinone. Catalysis. Works of republic conference on catalysis, Almaty, 1998, pp. 37-52.

Heaven, S. and Ilyushchenko, M.A. et al. (2000) Mercury in the River Nura and its floodplain, Central Kazakhstan:I. River sediments and water. The Science of the Total Environment, 2000, v.260, pp.35-44.

Nechaev E.A. (1989) Chemosorption of organic compounds on metal oxides. Higher School, Kharkov, p. 145

Panteleev V.G., Melentiev, V.A. et al (1978). Ash-Slug materials and ash lagoons. Moscow: Energy, p.295

Priimak T.I. and Timashev, V.V. (1982) About polyfunctional properties of slug-silicate sorbent. Silicate materials from mineral raw materials and industrial wastes. Leningrad: Science, p.372

Yanin E.P. (1997) Mercury in surroundings of the City of Temirtau . Central Kazakhstan, Moscow,p.30.

CHAPTER 24
Briquetting of Mining and Metallurgical Wastes

Dr. Valentin Aleksandrovich Noskov
The Iron and Steel Institute,
Academy of Sciences (Ukraine),
ul. Chernyshevskogo, 11A, kv. 20
49005, Dnepropetrovsk
Ukraine

Prof. Vadim Ivanovich Bolshakov
The Iron and Steel Institute
Academy of Sciences (Ukraine)
pr. Gagarina, 104, kv. 21
49107, Dnepropetrovsk
Ukraine

Summary

The main iron-containing metallurgical wastes and the possibility of their utilization have been studied. It has been shown that their downstream metallurgical usage is only possible after their preliminary treatment, primarily, by lumping. Priority has been given to briquetting as one of the state-of-the art methods of lumping. The work describes its features and advantages, shows the interest of enterprises in the usage of briquetted technogenous wastes in metallurgical processing. There have been studied the reasons of the slow commercialization of the briquetting process, and formulated the tasks to upgrade the briquetting technologies and enhance their usage.

Introduction

In the key industries of Ukraine - ore and coal mining, metallurgical, mechanical engineering and chemical engineering, with every passing year enormous volumes of solid wastes have been accumulated that can be utilized after an adequate processing, while their major part, in spite of their value as raw materials, present an environmental hazard. Therefore a solution of the problem of rational utilization of technogenous wastes is of utmost importance. This is dictated by the following: first, in the accumulated masses of technogenous wastes the content of the useful components sometimes exceeds their volume in the available or newly developed deposits, and, second, the accumulation of technogenous wastes adversely affects the environmental situation not only in the given region, but in the whole state.

The return of those wastes into the industrial production cycle is a key issue of the rational utilization of mineral and energy resources, production of reusable resources, development of non-waste and low-residue technologies, environment improvement in industrial regions.

Due to various reasons not all wastes are reused in the metal production, but are stored in dumps and slime ponds, as a result there are losses of raw materials, pollution of the environment, waste of land territories for their storage.

A study of the structure and chemical composition of technogenous wastes has shown that metallurgy is among the key industries where technogenous wastes or abundant resources can be

W. Leal Filho and I. Butorina (eds.),
Approaches to Handling Environmental Problems in the Mining and Metallurgical Regions, 247–256.
© 2003 *Kluwer Academic Publishers. Printed in the Netherlands.*

reused, including coke breeze, iron ore material dusts, slimes, cinder and off-take powder-like materials that are continuously generated in the production processes.

2 billion mt of wastes with an up to 36 % iron content were accumulated in the slime storage ponds of the six high capacity beneficiation plants of Ukraine, while the iron-containing wastes (ICW) of the iron and steel works have even a higher iron content.
At the full production output of the iron and steel works in Ukraine the annual ICW yield exceeded 10 million mt, including 4.0 million mt of slimes, above 6 million mt of solid wastes (blast furnace dust, cinder, welding slag, sinter and pellet dust, etc.).

The total amount of dispersed residues of metallurgical production per 1 mt of steel: specific yield of dust – 100-200 kg, slimes – 60-80 kg.

The finely dispersed materials include not only the residues, but also iron ore and in some part – manganese concentrates, as it is necessary to grind them during the beneficiation process. Though the finely dispersed basic raw materials for metallurgy are readily lumped using agglomeration and pelletizing methods, in some cases it becomes reasonable to take into account the possible use of briquetting technologies for their lumping.

The key criteria for the utilization fitness of metallurgical wastes are their chemical composition and humidity that determine their flowability and transportability. The main ICW materials with a 33-74 % iron content are dusts, slimes and cinders. Their sources: sintering, iron-making, steel-making and rolling facilities. Besides, these wastes feature a higher, in comparison to a starting raw material, content of calcium, manganese and carbon oxides [1].

But the presence of deleterious impurities in the bulk of the metallurgical dusts and slimes in the form of zinc, lead and alkali metal oxides caused the currently existing disproportion between the ICW generation and utilization.

A study of literature data has shown that to obtain high strength products (pellets and briquettes) the humidity of the slimes during pelletizing should not exceed 10-15 %. An optimum way to reduce the humidity of slimes from the accumulation ponds (20-40 % humidity) is by mixing them with a dry blast furnace dust, lime and other binders, while the mixing ratio of the components is selected taking into regard their subsequent processing and usage [2,3].

Both in Ukraine and other countries the iron containing wastes are mainly used as additives to the main burden materials in the production of sinter, and in a lesser amount, of pellets.

Along with this, lumping of wastes and fine fraction raw materials by briquetting has been started during the recent time. The technology of briquetting is quite diversified. Depending on the materials being processed and the requirements to the lumped product, cold and hot briquetting both with a binder or without a binder is used. Briquetting allows producing a lumped material of a specified shape and dimensions, density and mass. It allows lumping of materials that are both homogenious and inhomogenious in their composition, structure and particle size, while keeping constant or additionally introducing carbonaceous and other useful additives and thus forming an optimum chemical composition of a raw material. In certain cases briquetting is not just a reasonable, but the only possible method of lumping. Among the advantages of briquetting is its effectiveness in small tonnage production facilities. An analysis of the available technical information has shown that in some cases the prime cost and the specific capital expenditure of briquetting various materials are lower than when sinter and pellets are made of them [4-6].

The ICW processing technologies that are being developed have to correspond to the methods of their subsequent usage. During the recent years there is a trend observed both in this country and abroad to use wastes in the same industry where they are generated. Therefore an optimum site for their processing is the same production facility. High investments are required for the construction of installations to process high ICW tonnages from different production facilities. It should be noted that in Ukraine (Donetsk) there is a single specialty briquetting facility "Donetskaya" that produces coal briquettes (as domestic fuel) with a total annual production capacity of 1 million mt. During the recent years the open joint-stock company "Briket" that owns the above facility has dealt with the issues of upgrading the technological potential of the facility in cooperation with a number of organizations including our Iron and Steel Institute (ISI). So, for example, a 10000 mt batch of coke breeze briquettes produced at the facility in 1997 was successfully tested at the "Azovstal" Iron and Steel Works and received a high mark of the metallurgists. The tonnage coke - briquettes replacement ratio was 0.85, and due to a substantial price difference between briquettes and coke (approximately 30-35 % cheaper) it resulted in a substantial reduction of the hot metal prime cost - to 3 USD per ton. Today there is a possibility to use the facility for the production of coke briquettes for the metallurgical industry of Ukraine, and coke residue briquettes for domestic needs and power generation.

Let us take under consideration the main types of metallurgical industry wastes that contain iron, carbon, manganese and other value adding components, as regards their preparation for usage in metallurgical processes.

Sintering slimes have coarser particles in comparison to blast furnace slimes that mainly contain minus 0.15 mm particles. The chemical composition of the sintering plant and blast furnace plant slimes is a function of the quality and type of the burden materials used. The blast furnace slime contains magnetite, hematite, quartz, coke, calcium ferrite, etc. In most cases the sintering and the blast furnace slimes flow into the ir common settling facilities, thus their composition is averaged. The main difficulty in the preparation of slimes for lumping is presented by their high humidity. Preliminary to their transportation to the sinter plant the slimes taken from the settling ponds are soaked on the dewatering plates till their humidity allows their handling and mixing with other sinter burden components [7]. The humidity of the slimes during briquetting can be reduced either by using lime-based binders or by their preliminary drying in drying units.

The blast furnace dust is a mechanical mixture of sinter, pellet, limestone, coke particles featuring a different granulometry and chemical composition. Its iron content is within the 40-55 % range. The mineralogical composition of the dust corresponds to the composition of the burden materials and is characterized by the same components as the blast furnace slime. 80 % of the blast furnace slag is comprised of minus 2 mm particles. The finest dust (5-25 μm) is comprised of pellet fracture particles, while the sinter and coke particles are coarser (0.12-0.45 mm in the average).

The blast furnace dust is a dry granular material that does not require any preliminary treatment before lumping. Therefore it is actually completely used in sinter production. The dust characteristics are also compatible with the briquetting process, but preliminary tests have shown that it is briquetted worse than the slimes, so it seems more reasonable to utilize it as a component of complex briquetting burdens.

Steelmaking slimes are finely dispersed and are mainly comprised of minus 0.05 mm particles. The phase composition of an open hearth furnace slime is comprised of maggemite, magnetite, hematite, vustite, quartz, slag. The BOF slag is similar to the open hearth furnace slag in its mineral composition.

In their chemical composition and size the steelmaking slags are perspective for the production of briquettes to be used in the blast furnace and steelmaking processes.

The main problem in the production of briquettes from steelmaking slimes, as well as from blast furnace slimes, is the necessity to dry them. Besides, removal of harmful impurities is required in some cases. The advantage of using the briquetting technology for lumping steelmaking slimes is supported by the fact that only this method allows to introduce up to 40 % of carbon into the lumped material that allows to consider it as a material to replace the metal charge in steelmaking converters.

Rolling mill scale is the richest ICW in the iron content. Depending on the steel being processed it can contain various valuable alloying additives. 95 % of the primary scale is comprised of minus 2 mm particles while the average particle size is about 0.5 mm. The secondary scale is more dispersive and contains 8-12 % of oils.

The primary rolling mill scale is fully utilized at metallurgical enterprises, but not always it is used rationally. Taking into regard that in some cases its iron containing part is mostly in the form of iron monoxide it is preferable to eliminate any oxidizing processes during its lumping.. from this viewpoint the briquetting process is very advantageous, especially for alloy steel scale as it is possible to obtain a lumped material of a valuable composition that is more reasonable to be used not in the blast furnace, but in the steelmaking process.

Briquetting of the secondary scale is made difficult due to the presence of oils, therefore a preliminary de-oiling of the scale is required. This presents a separate technical task. If the secondary scale is used as an additive to the complex burden the former can be used without any de-oiling procedure.

Coke breeze (0-10 mm) is generated during coke sorting at coke chemical plants and during its screening upstream of the blast furnaces. The coke breeze from the Donbass coals contains 18-23 % ash, 13-16 % moisture, 5-6 % volatile matter and 1.7-1.8 % sulphur and has a lower reactivity.

At the present time at those iron and steel works which have sintering facilities, about 60-70 % of the coke brees is used for sintering while the rest is either stored thus complicating the inherently difficult situation at the ore yards, or is sold to outside customers on a limited scale. The situation is still more difficult at the works that have no sintering facilities, and at the coke chemical facilities. In the course of time storage usually results in the deterioration of the coke breeze quality and of the local environmental conditions. Its transportation to consumers in the unprepared condition is not economically viable for two reasons: high transportation costs and inadequacy for subsequent usage. At the same time its fractional composition is quite favorable for upgrading its commercial and technological characteristics by briquetting. Trial tests proved that the briquettes made of coke breeze can be used not only as a fuel, but also as a technological material in the blast furnace and cupola smelting.

Products of slag dump processing during the recent years have aroused an increasing interest at a number of metallurgical enterprises and firms cooperating with them. The steelmaking slags contain a substantial quantity of iron, either in the form of separate metal prills or other formations. To separate the metal containing material installations of various complexity and capacity are used, while their key components are magnetic separators of various designs.

The coarse fraction can be directly used in metallurgical installations while the 0-5 mm fraction has to be lumped. Half of its iron content (up to 60 %) is comprised of the metallic phase. In this case briquetting is the best lumping alternative as it allows preserving the metal in the non-oxidized condition. As regards their granulometry and humidity, the fine fractions of the metal burden produced by slag processing are fully prepared for briquetting with binders

The possibility to lump fine fractions of metal burden is of importance for all industrial facilities where slag dumps are processed. Therefore the development of technologies and equipment for lumping of metal-containing fines is of high importance.

Ferroalloy production wastes are generated at all stages of charge materials preparation and smelting of ferroalloys. Ore fines and dust, screenings of manganese concentrates and carbonaceous reductants are formed during the screening of materials, inter- and intra-shop handling and transportation. Their chemical composition is actually identical to the respective ores, manganese concentrate and coke. The size of the ore fines is minus 3 mm, coke screenings - minus 5 mm. During the ferroalloys smelting powder-like residues are formed - slimes, gas cleaning and aspiration systems dust, fine fraction slag residues, fines from ferroalloy crushing. All those materials cannot be utilized in the smelting processes without their preliminary lumping. Briquetting as a method of lumping fine fraction materials is more used in the ferroalloy production than in other metallurgical processes, and it proved to be effective in a series of cases.

The Iron and Steel Institute (ISI) has studied all types of wastes generated at the Nikopol Ferroalloy Plant (NFAP), Zaporozhye Ferroalloy Plant (ZFAP) and Stakhanov Ferroalloy Plant (SFAP) to determine the possibility and feasibility of their briquetting. The main criteria of the analysis: chemical composition, granulometric composition, quantity, conditions of collection, storage and transportation, alternative lumping methods and their inappropriateness.

At the present time the ISI has been developing a technology and pressing equipment for briquetting of silicomanganese and ferromanganese screenings generated at the NFAP and "Krivorozhstal" Iron and Steel Works.

Metal chips and turnings of various sizes are in part generated at metallurgical enterprises, but mainly at mechanical engineering and other metal processing enterprises. The volume, product range, ways of the materials utilization are being evaluated now.

The bulk is comprised of the steel, iron and aluminium chips and turnings. The coarse steel and iron materials are mostly prepared for their further processing at the enterprises of secondary ferrous materials. But in some cases, especially as regards high alloy iron and steel products, it is more reasonable not to mix such chips and turnings with other materials, but to prepare them for direct usage in smelting of respective steel and iron grades. For instance, at the Dnepropetrovsk rolls manufacturing plant the issue was studied of briquetting alloy iron chips for in-plant usage. The same issues were discussed at the "Dneprotiazhmash" plant (metallurgical equipment producer) and at the Lutugino rolls manufacturing plant (Russia).

The generated aluminium turnings are fully utilized according to a variety of technologies.

Steelmakers repeatedly raised the problem of briquetting aluminium tirnings to be used in steelmaking. In such case a product of a complex composition can be produced, for example, by adding chips of heavier metals that is important for the technological process performance. Such residues as lime screenings and lime dust also require some attention.

Almost all integrated iron and steel works have shops or plants for limestone calcining which prepare lime mainly for the oxygen converter shops. The screenings of the burnt lime below 10 mm in size are not fed directly to the metallurgical units though in some cases they contain more free (active) calcium oxide than the lumps. The finely dispersed lime dust is caught in large volumes in the dust removal systems of the calcining units. As dust is generated at various production stages, including the unburned material grinding stage, the CaO_{free} content in the dust is unstable and less than in the screenings. Presently the lime dust is mostly used at the plants for pre-drying of slimes before they are fed into the sinter burden. At some foreign enterprises there were attempts of lime screenings briquetting for downstream usage instead of lumped lime.

Lime residues should be considered as a component of briquetting burdens that is not only useful during the metallurgical processing of the briquettes, but also serves as a binder in their manufacturing. Lime screenings are more suitable as a binder due to their composition, but they require further grinding. The effectiveness of the lime dust as a binder is lower (due to a lower content of the active calcium oxide), but its fraction composition does not need any adjustment.

It is worth noting that the usage of lime residues as a component of the briquetting burden allows, due to its dewatering effect, briquetting of materials (such as slimes) with a higher humidity.

Thus, the above discussed materials show that among the wastes, in particular iron-containing wastes, there is a large volume of fine fraction and dust-like materials that require preparation, mainly lumping, for downstream metallurgical processing.

As regards the interest displayed by enterprises in utilizing wastes in the briquetted form it should be noted that the main criteria are the high technological properties of the briquettes and their lower cost in comparison to sinter, pellets, coal, coke, etc.

The main tasks set to achieve the above and to enhance usage of the briquetting technologies and their effectiveness:
> development of a strategy for an integrated processing of industrial wastes;
> production of briquetting equipment with a wide range of technological and power parameters for processing of wastes with varying characteristics;
> improvement of methods and means for dewatering of materials preliminary to briquetting;
> development of novel binding additives;
> provision of the specified technological properties of the briquettes.

The developments in briquetting that took place in the home and foreign metallurgical industry from the 50th till late 80th of the XX century were highlighted in the works by L. P. Eidelman and B. M. Ravich [4,5]. It was shown that the results of the commercial and pilot scale smeltiong operations with briquetted burdens in the actual conditions of metallurgical enterprises demonstrated a number of advantages of using briquetted burdens:

- improved conditions of the lumped material drying, transportation and reloading;
- increased bulk density of the briquetted burden;
- increased gas permeability of the briquetted burden layer;
- increased briquette value for the consumer due to the dimension and shape uniformity;
- reduced losses with the emissions during preparation and usage in the technological processes;
- reduced susceptibility to caking in storage, and to lumping during usage, to arching during transportation through discharge chutes, and to hovering in bins;
- reduced oxidation rate and decomposition rate during lengthy storage (due to the high material density in the briquette surface).

The briquetting issues have become very acute in the home and foreign metallurgy during the latest decade due to the depletion of deposits, severe competition in the ferrous metal market, continually increasing technogenous wastes and stricter environmental requirements. Briquetting is most advantageously used in the metallurgical industries of the USA, Great Britain, Germany, Poland, South Korea. Japan, France. Interest is displayed to this problem in China, India, Turkey.

During the recent years briquetting has become an important issue in Ukraine due to the adoption of the "Law on Wastes", the "Concept of Mining and metallurgy development in Ukraine till 2010", and the state "Program of Production wastes Utilization and Usage till 2005".

The slow commercialization of briquetting is explained by the lack of home production of presses on a commercial scale. The issue of designing and manufacturing of reliable home made presses for briquetting of fine fraction raw materials and industrial wastes has become of primary importance in the recent years. It is explained not only by the need to process technogenous wastes, but also by the fact that Ukraine and other CIS countries had not any experience in the design and manufacture of roll type briquetting presses. The available inadequate quantity of roller type presses produced in this country and operating at a few enterprises were developed and manufactured by various enterprises specialized in other industrial fields. Up to now there has been no unified methodological approach to the principles of calculation and designing of briquetting presses as a whole and of their main units. The presses have been developed and operated without due consideration of the burden materials properties and specific features of their briquetting technologies. This leads to a discrepancy between the design and actual loads that develop in the main units and in the drive, and subsequently to a reduction in the product quality.

A certain amount of experience in the designing and manufacturing of briquetting presses has been accumulated in Germany, France, USA and Poland. Their purchase for the enterprises in this country meets numerous difficulties, first of all, in selecting a proper press for the given material to be pressed, and for the given shape and dimension of the briquette. Besides the high cost of the presses and spare parts requires substantial hard currency expenses.

To extensively introduce briquetting in our industries it is necessary to produce pressing equipment in this country on the basis of a scientifically substantiated approach.

In the production of briquettes from fine fraction technogenous wastes and raw materials the granular materials have to be transformed to solid bodies. During the briquetting process the granular burden is found in different states: in the condition of a relative temporary immobility in the collecting bins, in the state of mobility in pipes and burden channels, while the density of the material is gradually increasing in the deformation zone of the rolls that catch the material.

The Iron and Steel Institute (ISI) has studied the behavior of loose burdens under the above conditions while modifying their physico-mechanical properties: this is one of the key issues of the briquetting process theory. Special installations and methodologies were developed to study the burden compaction process in a closed volume. Graphical and analytical relationships between the burden compaction and the compression pressure were derived. Burden binding additives were selected and the external and internal friction characteristics of the tested burdens were determined. Burden briquetting schedules were finalized in experimental roll type presses while studying the major briquetting parameters. On the basis of the studies that had been carried out, technological schemes of the briquetting process and a scheme of the apparatus chain were developed for various burden materials; designs of the pressing equipment were studied and new technical solutions proposed.

The undertaken theoretical and experimental studies allowed to develop a methodology and an algorithm for calculating technological, geometrical and power parameters of the burden briquetting process in roll type presses. In essence, new issues of the briquetting theory were developed which allow to select and calculate the design parameters of roll type presses on a well grounded basis, taking into regard the technological requirements and the commercial properties of the briquettes.

Using the methodology and the algorithm of parameters calculation, a unique design of roll type presses has been developed featuring simplicity, operating convenience, reliability, relatively small dimensions, repair ability and reduced metal consumption. Their technical characteristics are given below [8]:

Technical characteristics of presses

* Capacity, t/h	0.5...20
* Press type	roll type
* Pressure, MPa, max.	120
* Compressing force, kN, max.	1000
* Drive power, kW, max.	30
• Rolls:	
diameter, mm, max.	800
working width, mm, max.	500
rotational speed, $\min^{-1}$	0.5...12
• Briquette:	
shape	according to customer's order
volume, cm^3, max.	100
* Press dimensions,	
LxBxH, mm, max.	2500x1900x1300
* Press mass, kg	4000...4500

The presses are manufactured at the ISI experimental and production facility. The presses were commercially tested and operated at a number of enterprises in Ukraine and Russia for briquetting of wastes at metallurgical, coke chemical and refractory plants, and of coal industry wastes.

The development of the briquetting technologies at the ISI is carried out over the whole cycle - from the laboratory studies of the burden intended for briquetting to the setting up a technological schedule for the briquetting facility design, for the development and manufacture of roller type briquetting presses featuring the required technical characteristics.

The technology of briquetting of technogenous wastes is implemented in accordance with a technological scheme of the briquetting process and a scheme of the chain of apparatuses. The production of briquettes includes various technological operations and equipment units.

When selecting and calculating the technological schemes for producing briquettes from fine fraction burdens, several factors have to be taken into account: the physico-mechanical characteristics of the initial burden components (humidity, looseness), their mechanical composition; requirements to the briquettes that are determined by the technology of their downstream processing (strength characteristics, moisture resistance, thermal resistance, etc.), the conditions of the delivery and storage of the burden components at the production facility, etc.

When selecting the scheme of the chain of apparatuses and of the equipment it is comprised of, the main parameters are: the required throughput of the briquetting facility, the ability of the equipment to provide the required technological functions, its reliability.

As a rule, fine fraction burdens are briquetted using binding additives, their selection being one of the most important technology elements. The binder is to provide the required strength characteristics of the briquettes; they should not introduce deleterious components during the subsequent processing of the briquettes, be readily available and not too expensive.

In spite of a large variety of binders, only a limited number of them are usable for high-tonnage production.

Taking into regard the available experience in the practice of briquetting various burden materials, as well as the literature data, the mostly wide usage in the briquetting, in particular, of metallurgical wastes have such binders as lime, liquid glass, technical lignosulphonate, molasses and their mixtures.

In most cases the briquetting technology includes the following operations: batching of the burden components, mixing, briquetting, screening, strengthening.

The main unit in the scheme of the chain of apparatuses is the briquetting press that is designed for the required throughput of the facility according to the properties of the given burden. Ether one or several presses can be installed to meet the required capacity. The rest of the technological and auxiliary equipment, such as mixers, screens, batchers, conveyors, etc. are selected from the commercially produced equipment. If a drying unit is envisaged in the scheme, it is also often manufactured according to an individual project.

Conclusions

The experience of industrially advanced foreign countries (USA, Germany, France, Japan, etc.) and the operating practice of certain enterprises in this country point to the perspectiveness and economic viability of briquetting in the ferrous metallurgy to be used for the preparation of fine fraction burden materials for smelting and for the utilization of metal-containing production wastes.

Along with the shown interest of metallurgical enterprises in the waste utilization by briquetting it should be noted that the main interest criteria will be the high commercial properties of the briquettes, etc. One of the main tasks in achieving these characteristics and in expanding the range of the briquetting technologies usage and effectiveness is the production of the briquetting equipment for processing of wastes featuring different properties and having a wide range of technological and power parameters. At present there are favorable conditions and possibilities for extensive implementation of the briquetting technology for lumping of raw materials and industrial wastes in various industries.

References

Savitskaya L.I. Utilization of iron containing wastes in lumping of ores. Inform. review // Ser. Preparation of raw materials for metallurgical processing and iron production/ Institute "Chermetinformatsiya". 1984.Issue 5. 27 p. (In Rusian)

Gubanov V.I., Komorzhenov G.I., Stepanov B.Ya. Experience in upgrading the technology of agglomeration and utilization of iron-containing wastes. Express-inform. Chernaya Metallurgiya. Institute "Chermetinformatsiya". 1982. (In Russian)

Korzh A.T., Golubov A.F. //Chernaya Metallurgiya. 1991. No. 5. P. 49-50. (In Russian)
Ravich B.M. Briquetting of ores. M.: Nedra. 1982. 182 p. (In Russian)

Eidelman L.P. Briquetting equipment and technology in the home and foreign ferrous metallurgy // Chernaya Metallurgiya. Bull. Inst. "Chermetinformatsiya". 1988. No. 8. P. 2-12. (In Russian)

Noskov V.A. Metallurg. i gornorud. promyshlennost. 1998. No. 3. P. 119-121. (In Russian)

Veletsky R.K., Kanenko G.M., Sakovsky V.D. Development of technology for processing iron-containing wastes of metallurgical enterprises in Ukraine. Vestnik Ukrainskogo Doma ekonomicheskikh i nauchno-tekhnicheskikh znaniy. 1999. No. 4. P. 227-228. (In Russian)

Noskov V.A., Bolshakov V.I. Recycling of metallurgical industry wastes by their briquetting. Proceedings, International Seminar on recycling of ferrous metallurgy wastes. UNO Economic Commission. Linz. Austria. - 1998. - P. 1-6. (In Russian and English)

CHAPTER 25
Neutralizing hazardous liquid waste in the coke industry

Dr. Victor Mikhailovsky

Enterprise Envitec,
4 Osvity str., Kiev 252037
Ukraine

Summary

This chapter presents the results of a study on liquid phase neutralizing research for the collecting pond of liquid wastes from the chemical plant "Marcochem". The volume of waste volume is about 80,000 m^3. The collecting pond is located in the clay lock-proof channel at a distance of up to 500 meters from the town of Sartana. Underground infiltration, pollution of the drinking water well and gas emission in the atmosphere were all registered. The main contaminations are as follows: phenols, ammonia and pyridine compounds, sulphates, solid residual up to 35 g/l. pH of medium is 1,5 =?a medium pH of 1.5. State of the art methods of neutralizing surface water with a low pH, for example by lime transportation on the surface of the ice in winter as in Scandinavian countries, are unacceptable in this case because an abrupt increase in pH will cause an intensive discharge of ammonia and pyridine. So, technology must provide a means of controlled gradual increase of the pH value. As a result of research, two stages - neutralizing and biological treatment - have been proposed. Liquid phase after biological treatment is expected to leave in the existing channel without discharge into the surface sources.

Contaminations of the liquid phase

The following major contaminants were determined in the initial liquid phase: phenols – 10.4 ppm; volatile ammonia – absent, combined ammonia – 8840 ppm; pyridine – 98 ppm; iron – 1200 ppm; sulphates – 23500 ppm; salinity – 35800 ppm; optical density – 0.42; pH – 1.5. Under normal conditions this liquid has the colour of tea and a characteristic phenol smell. Value =?the volume of liquid phase in the collecting pond is about 80 000 м^3. The low pH-value is due to the essential content of free sulphuric acid H_2SO_4. The main contaminants are characterized as follows:

Phenols, and their derivations such as cresol, xylenol, thymol are the weak acids. In the strong-acid medium they are in an uncombined state and particularly transform into the gaseous phase which causes the phenol smell of the liquid phase. As a result, the liquid phase is in a state of unstable equilibrium with the gas phase. Simultaneously, phenols diffuse from tarring to the liquid phase. Consequently, atmospheric pollution by phenols takes place permanently with the fluctuations in concentration depending on the ambient conditions (temperature fluctuation, wind, precipitation etc.) and is on average 0.2 ppm close to the liquid phase surface. Phenols also diffuse into subsoil waters polluting it. The almost permanent =?constant level of the liquid in the collecting pond is in equilibrium with the level of subsoil waters.

Ammonia, a weak base, ($K_д$=1.8 10^{-5}) at pH = 1.5 and H_2SO_4 sulphuric acid excess is in a fully combined state in the form of ammonia sulphate and ions of ammonia. The compound is

W. Leal Filho and I. Butorina (eds.),
Approaches to Handling Environmental Problems in the Mining and Metallurgical Regions, 257–265.
© 2003 *Kluwer Academic Publishers. Printed in the Netherlands.*

sufficiently stable at this pH value. Therefore, emission of ammonia is practically absent. The presence of uncombined ammonia in the liquid phase was not detected.

Iron is completely combined in the form of good-soluble iron (III) sulphate. The tea colour of the liquid phase is also due to the presence of iron ions (III) in the initial liquid phase.

Pyridine, a weak base ($K_{д}$=1.5 10^{-9}), is in a completely combined state in the form of pyridine sulphate. However, this compound is not so stable as ammonia sulphate and exists only in an acid medium where there is an excess of sulphuric acid H_2SO_4. Pyridine sulphate is destroyed as a result of the hydrolysis in a neutral and weak-base medium. The collecting pond and the process of environmental pollution are schematised in Fig.1

Fig.1. Influence of a liquid waste collecting pond on the environment

Pilot research on the neutralization process and biological treatment of the liquid phase

A pilot installation for the neutralization process and biological treatment of the liquid phase has been installed on the industrial site of JSC «Marcochem» at the biological treatment plant. The installation consists of a metal cylindrical body (1) with a deflection nipple, which is connected with the circulation pump (2). The regulator (3) controls flow velocity.

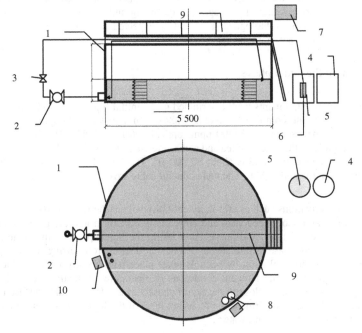

Fig. 2. Pilot installation

1 - frame; 2 - circulation pump; 3 - regulator; 4 – slacker tank; 5 - lime milk tank; 6 - sampler of lime milk; 7 – control panel; 8 – gas analyser; 9 – bridge for service; 10 – pH-meter

Liquid feed has a specific direction in order to create a circulation flow. The flow is distributed over the tank cross-section, providing effective mixing of lime milk in the volume of the liquid phase. Lime was slaked in tank 4, after which a slake lime was diluted to the require concentration (10%) in tank 5. Lime milk from tank 5 was fed into the suction line of the circulation pump by the sampler (6). The circulation pump also acted as a mixer.

Neutralization of the liquid phase

The neutralization process was spread over 4 days. This period was determined by the requirement that the pH should be raised gradually in order to prevent gas emission to the atmosphere. Measurements were made to determine the solution content in the test installation and air content at a distance of 10-15 cm from the surface. The results of the experiment are shown in Tables 1 and 2.

Table 1.

Liquid phase, ppm		Gas in air, mg/m^3	
Phenol	10.4	Phenol	Trace
Ammonia volatile	absent	Pyridine	absent
Ammonia total	8840	Ammonia	absent
Pyridine	98		
PH	1.5		
Sulphate	23500		
Optical density	0.42		
Salinity	35 800		

Table 2

Liquid phase, ppm		Gas in air, mg/m^3	
Phenol	3.3	Phenol	absent
Ammonia volatile	180	Pyridine	0.2
Ammonia total	4180	Ammonia	10
Pyridine	44		
PH	7.5		
Sulphate	13 400		
Optical density	0.14		
Salinity	13 160		

Fig.3. Neutralization process phases
a – initial state of solution, в – in the mixing process, c – after suspension sedimentation

The initial state of the solution (Fig.3 a) corresponds to the liquid phase of a sludge settling tank. The liquid has dark brown colour and smells of phenol. The gradual increase of pH as a result of the lime milk dosing, leads to a suspended phase formation consisting mainly of gypsum ($CaSO_4$) and metal hydroxides. Ammonia, pyridine and phenol compounds settle primarily on the active surface of metal hydroxides and gypsum crystals (Fig.3в). The formed suspension settled rapidly. Bound between residual and clarified water is observed after a 12-hour settling period (Fig.3c). Contaminant concentration in the solution is significantly reduced due to particular contaminant sedimentation. The results of the sedimentation are shown in Table 3. pH adjustment and a control system was used throughout the research into the neutralization process.

Table 3

Liquid phase, ppm		Gas in air, mg/m^3	
Phenol	0.14	Phenol	absent
ammonia volatile	82	Pyridine	traces
Ammonia total	720	Ammonia	4
Pyridine	21		
PH	7.7		
Sulphate	12700		
Optical density	0.06		
Salinity	13 100		

Neutralization was carried out by dosing 10% lime milk in the liquid phase. The process was carried out so that the pH value increased evenly. When dosing lime milk the use of even a diluted solution of H_2SO_4 is undesirable because of the formation of a gypsum layer on the $Ca(OH)_2$ surface. This layer makes the dissolution of calcium hydroxide difficult. Besides, pyridine and ammonia will emit intensively during lime milk preparation. However, in the process of gradual neutralizing, the influence of undesirable factors will be eliminated, due to the following:

The contact time of the suspension $Ca(OH)_2$ with the liquid phase that must be neutralised is long enough at the intensive mixing in the turbulent flow. It allows all the lime to react and emitted pyridine lesser pH gradient over the pond will be kept. Hence, intensive emission in the atmosphere is eliminated.

When neutralizing we see the following:

pH 1,5 – 3. Only neutralizing of H_2SO_4 takes place. At the same time gypsum precipitates:

$$H_2SO_4 + Ca(OH)_2 = 2 H_2O + Ca SO_4$$

Small local increase of pH in the microvolume near dissolving calcium hydroxide is the cause of a slight increase in pyridine concentration in the gas phase.

$$C_5H_5N\text{-}HSO_4 + CaOH^+ = C_5H_5N + H_2O + Ca SO_4$$

pH 3 – 7. Calcium ions Ca^{2+} forms with sulphate-ions bad-soluble compound (gypsum) leading to the destruction of the soluble sulphates:

$$(NH_4)_2 SO_4 + Ca(OH)_2 = Ca SO_4 + 2 NH_4OH$$
$$NH_4OH = NH_3 + H_2O$$
$$Ca(OH)_2 = CaOH^+ + OH^-$$
$$CaOH^+ = Ca^{2+} + OH^-$$
$$C_5H_5N\text{-}HSO_4 + CaOH^+ = C_5H_5N + H_2O + Ca SO_4$$

Emitting pyridine and ammonia are absorbed by the liquid phase and particularly emit in the atmosphere. The presence of the above components gives the medium to be neutralised the reduction characteristics. Reduction process from Fe^{3+} to Fe^{2+} takes place and the medium changes colour from brown to green. Fe^{2+} spontaneously passes into Fe^{3+} contacting with the oxygen in the air. The iron concentration in the liquid phase of the collecting pond is 0,02 g-mol/L. Hence, it is possible that iron ions content in the liquid phase will be limited by oxidation velocity of Fe^{3+} to Fe^{2+}. Sedimentation of Fe^{3+} in the form of $Fe(OH)_3$ takes place at pH=3...5. A gradual neutralization process to avoid reduction from Fe^{3+} to Fe^{2+} leads to the iron sedimentation in the form of iron hydroxide (III) even at this stage.

pH 7 – 8. Pyridine and ammonia emission is more intensive than at the previous stage because of the dissociation due to an additional injection of hydroxides ions with $Ca(OH)_2$ solution in the system. Equilibrium in the systems

$$NH_4^+ + OH^- = NH_4OH = NH_3 + H_2O$$

and

$$C_5H_5NH^+ + OH^- = C_5H_5NH(OH) = C_5H_5N + H_2O$$

accents in the side of uncombined pyridine and ammonia more and more. Pyridine and ammonia concentration in the gas phase rise and at the same time, even at a pH of 8 this concentration does not reach the tolerance limits for working areas. Air was sampled near the liquid phase surface. Formation of $Fe(OH)_2$ residual starts at pH 7,5. Hence lime milk should be dosed gradually to prevent intensive emission of ammonia and pyridine.

Biological treatment of liquid phase

In the pilot investigation a biofilter loaded with coke was used. (Fig.4). The biofilter in this case acted as a bacteria incubator. Populations of bacteria growing on the filter bed were circulated all over the tank. Liquid was pumped into the distribution system. The biofilter was charged by a crop from the biological treatment plant of JSC "Marcochem". The circulating flow ranged from 0.5 to 0.8 m^3/h.

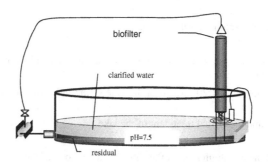

Fig.4. Pilot installation for biological treatment

Since the average air temperature had fallen below 6°C and the progress of the biological process was slow, tests on biological treatment were continued in the laboratory. The laboratory installation is shown schematically in Fig.5. The tests were carried out in the Laboratory of Water Biological Treatment Methods in the Institute of Biology at the Ukrainian National Academy of Science.

A pipe 1500 mm high and 70 mm in diameter was used. Wheat straw stems were used as the filter bed. The filter was charged by a crop adapted specially for the treated solution by the Institute of

262

Biology. Laboratory research confirmed the efficiency of the biological treatment in that the concentration of impurities diminished significantly. It is expected that the biocenosis will work even more effectively in the full scale stage of implementation, increasing the efficiency of the treatment.

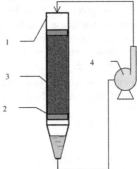

Fig.5. Scheme of lab installation for biological treatment
1 – biofilter body; 2 –frame; 3 – filter bed; 4 – circulation pump

Technical solutions

The concept developed to solve the problem was based on the following:

- Minimal emissions in the environment
- Minimal capital and operating expenses
- Wastes minimisation

Several methods were examined, for example:

- Building a treatment installation on the bank of the lake and following discharge of the treated water into the river
- Building of a pipe (about 20 km) from the lake to the industrial site to treat wastewater at the existing installation
- Neutralization and biological treatment of the lake

The prevention of the contamination of the subsoil water is one of the aims of the project. We could hardly solve this problem using the first two variants as a fall of the water table in the lake will be compensated by the new subsoil waters and by precipitation. Hence, pumping water out of the lake could be an interminable project. Besides, the first and the second variants would cost about $ 1,500,000- 2,000,000 in capital investments.

Neither of these suggestions seems good enough to justify the expenditure involved. The alternative solution involving neutralization of the liquid phase, including its further biological treatment, seems more attractive. This variant involves the following:

- Neutralization of liquid phase by lime milk

• Adaptation of microorganisms to the neutralised solution

• Formation of biocenosis in the collecting pond

In order to achieve the above goals, a circle flow of the liquid phase is created, in which the lime milk is dosed. A dosing pump injects the lime milk in the circling flow with a definite controlled discharge to control gas emissions in the air. This process of neutralization of the liquid phase is shown in Fig.6

The solution flows through the water scoop (1) to the circling pump. The circulation pipeline acts as a mixer of the lime milk and the liquid phase. Lime milk is prepared in the tanks (3).

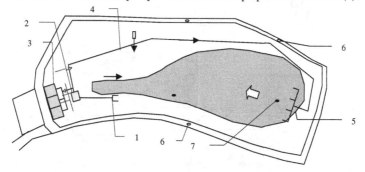

Fig.6 Scheme of the neutralization of the liquid phase

1 – water scoop, 2 – circulation pump, 3 – lime milk preparation tanks, 4 – circulation pipe, 5 – water-discharge, 6– gas analysers, 7– pH –meters

The neutralised liquid phase enters through the circulation pipe in the opposite side of the lake through the water-discharge(5) mixing with the solution in the lake. The gradient of pH in the zone of the water-discharge is very low and does not exceed 0.05 pH per 1m. So, the creation of local zones with higher concentration $Ca(OH)_2$ and zones with intensive gas emissions are prevented. The scoop of the neutralised liquid is located in the shallow part of the lake and the discharge installed on the opposite side of the pond. This will exclude the appearance of turbulent flows and create conditions for the even precipitation of colloid suspension.

Lime is one of the most effective and low price reagents for acid pollution neutralization. Furthermore, lime is the traditional component for acid soil conditioning and for combining and localizing such dangerous pollutions as heavy metal ions and radioactive waste. In this project, lime not only normalizes the pH value in the liquid phase of the collecting pond but also combines and transforms an essential part of the dissolved pollution from the lake in the colloidal phase. Besides, it is known that calcium compounds (carbonates and sulphates) tend to crystallize and cover the surface. This may be useful because calcites will plug up blowholes which are the cause of pollutant infiltration into subsoil waters at present.

Hazardous pollution of the environment by collecting ponds will diminish as time goes by, Fig.7

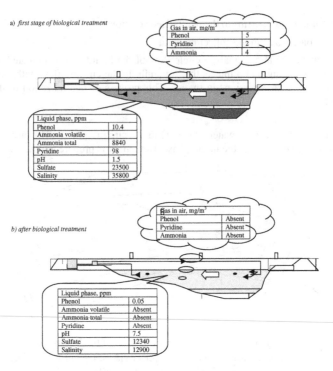

a) *first stage of biological treatment*

Gas in air, mg/m³	
Phenol	5
Pyridine	2
Ammonia	4

Liquid phase, ppm	
Phenol	10.4
Ammonia volatile	-
Ammonia total	8840
Pyridine	98
pH	1.5
Sulfate	23500
Salinity	35800

b) *after biological treatment*

Gas in air, mg/m³	
Phenol	Absent
Pyridine	Absent
Ammonia	Absent

Liquid phase, ppm	
Phenol	0.05
Ammonia volatile	Absent
Ammonia total	Absent
Pyridine	Absent
pH	7.5
Sulfate	12340
Salinity	12900

Fig. 7. Pond influence on the environment with time

The flocculation phase gradually reacts with tarring. The tarring layer will be covered with deposits and it is expected that the lime will form an insoluble crust at the bottom of the lake. The crust, in conjunction with suspended gypsum and calcium carbonate, forms a protective screen isolating the liquid phase of the lake from the subsoil water. Heavy metal salts and iron compounds also settle at the neutralization stage. Iron, particularly in the form of complex compounds, therefore speed of transformation in hydroxide and following settling is less than sedimentation speed of the common system. The complex compounds are destroyed in the process of biological treatment and iron concentration is reduced to 0.1-0.3 ppm. Ariil sulphonates are also destructed, but this process has been already investigated [1].

Conclusions

Pilot investigations have given a set of results. Firstly, even at the neutralization stage, concentrations of the main contaminants are reduced: phenols –69%, total ammonia –53%, pyridine –55%, sulphates –44%, salinity –43%. The pH value became 7.5 and optical density increased by 67% creating favourable conditions for biological treatment. Secondly, even at the first stage of biological treatment, the subsequent reduction in concentration of the main contaminants with respect to the initial concentrations is as follows: phenols – 98%, total ammonia – 92%, pyridine – 77%, sulphates – 46%, salinity – 43%, and optical density increased by 86%.

Discharges into the air from the surface of the initial solution consist mainly of phenol compounds. Pyridine and ammonia begin to evolve with the rise in the pH value. A preliminary analysis of the evolving process has shown that the concentration of these substances close to the surface did not exceed tolerance limits for working areas.

The technique developed solves one of the main ecological problems of the Donetsk industrial region. The problems caused by this type of pollution are severe enough at the present time, but it could cause an ecological disaster in the future when not only subterranean waters and the River Kalmius but also the coastal region of the Sea of Azov may be irreversibly polluted.

Technical decisions involving neutralization followed by biological treatment are environment-friendly and cost-effective.

References

Utkin E. B (1991) Destruction of toxic organic compound by microorganisms. Research Paper. Moscow.

CHAPTER 26

Refractory Materials and their Influence on the Ecological Situation in the Steel Industry
- Ecological and economic aspects

Dr. Vasyl Moskovchuk

RHI AG,
Wienerbergstrasse 11, A-1100 Vienna
Austria

Summary

This chapters describes the role refractory materials can play in the steel industry, with some considerations on the ecological and economic aspects.

Introduction

Mining and metallurgical complexes (GMK) take second place only to the power generating industry in terms of their contamination of the environment. Higher morbidity and mortality rates in the population and a falling birth rate can be observed in areas with a high concentration of GMK – plants. The extent of harmful pollution of the industrial regions of the Ukraine can be seen as not less dangerous, but possibly more dangerous than that caused by the Chernobyl tragedy (Table 1).

Table 1. Anthropogenic contamination of the environment by heavy industry in the Ukraine [1]

Industries	air pollution	water consumption	waste water
	1.000 t/a	million m³/a	million m³/a
TOTAL	2640	1786	1495
metallurgy	1406	1535	1292
mining	804	114	112
coke-chemical	182	41,8	15.5
tube rolling	85	31	24.2
refractories	69	6	9.7
ferro alloys	75.9	5.7	0.9
engineering	2.5	1.3	0.7

For Russia, with a 14.4% increase in metallurgy production and a 20.8% increase in the refractory industry (1999) [2], there was an increase of about 6.4% in harmful pollution (Table 2), caused in essence by the non-observance of technical requirements. According to the calculations for 1999, solid wastes composed about 15%, nitrogen oxides about 8%, hydrocarbons over 12%, carbon oxide 25%, and sulphurous anhydride 35% of the 14.7 million t of industrial pollution in Russia. For ferrous metallurgy, with 2.3 million tons of pollution, the corresponding figures were 14.5; 6; 0.3; 68.5 and 10%.

W. Leal Filho and I. Butorina (eds.),
Approaches to Handling Environmental Problems in the Mining and Metallurgical Regions, 267–280.
© 2003 *Kluwer Academic Publishers. Printed in the Netherlands.*

Let us compare this situation with ecology in metallurgy in Western Europe, based on the example of the refractory industry in Germany. The figures given in the monitoring report for 1999 [3] state that of approximately 3 million GJ/a (1998 data) in the consumption of primary energy resources, the refractory industry was responsible for a major proportion through natural gas (> 95%), which decreases the general rate of direct emission of CO_2 during the production of refractory materials (171.342t/a).

Table 2. Pollution development in air (1) and waste water (2), Russia [2]

Industries	(1) 1998	(1) 1999	(2) 1998	(2) 1999
	1000 t	1000 t	Million m^3	Million m^3
RUSSIAN FEDERATION	18661,82	18539,67	21986,18	20657,01
Industry, total	14949,82	14704,44	6867,89	6445,32
Energy, power	4345,69	3935,51	1448,14	995,20
Non-iron metallurgy	3291,79	3311,83	377,5	363,58
Steel industry	2188,94	2329,59	676,93	698,96
Oil mining	1385,03	1328,99	10,91	4,34
Coal mining	545,31	559,96	442,05	395,99
Natural gas	428,48	456,26	3,25	3,15
Engineering	60,05	454,10	552,33	596,84
Building materials	396,56	416,94	112,09	121,77
Chemical& petrochemical	388,02	414,93	1240,32	1248,98
Nuclear power	105,0	92,1	252,7	297,8

Indirect emission of CO_2 in the same period was 84.515 t (from the net 153.664 MWh /a). With a total volume of 1,088 million t of refractory materials, the overall energy consumption was 4,561 million GJ/a, or 4,19 GJ/t, whilst specific emission reached 0,235 t/t [3], including indirect and direct emission of CO_2 (255.857 t/a). Taking into account the target to reduce the emission of CO_2 by 15-20% for the refractory industry and by 16-17% for the metal industry (21-27 million t in 1990 and 17 million t forecast for 2005) [4], the role of refractories (80 enterprises with 8,800 workers and a turnover of 2.5 billion DM in 1996) seems to be insignificant (only 10%) in CO_2-emission from Germany's metal industry (1996: $570*10^9$ MJ of primary energy and $11*10^6$ MWh of electric power or $685*10^9$ MJ of total energy consumption for 31 million t of steel with 43 billion DM turnover and 2,2 t CO_2 /t of LS and predicted reduction to 1,734 t/t in 2005).

Nevertheless, modern refractory solutions have been more effective in reducing CO_2 pollution than even the best modernisation processes in the refractory industry itself. For example, the optimisation of the supply of secondary air in the bell furnace of one of the refractory plants decreased specific heat consumption from 1770 to 1275 kcal/kg, and CO_2-emission from 438 to 316 kg/t, i.e., by 28%. Other measures which should be mentioned here are: the installation of cooling towers to replace thermal post-combustion, the installation of an additional injection of air in the tunnel kiln and the shortening of drying time (400 °C) for the mouldings. All these changes have served to reduce emission; nevertheless, 10% specific pollution caused by refractories indicates that it is necessary to search for ecological improvement potentials in the steel industry where the application of refractory materials is involved. Therefore, the principle areas of ecological modernisation in the steel industry [5] include:

- More efficient prevention of slag carryover on tapping
- Improved consistency of metallurgical performance during long campaigns, bath agitation, etc.
- Improved durability of converter vessels, improved cooling systems to prevent cone distortion
- Reduced energy consumption and losses

- Extended service life of secondary steel making equipment and ladles
- Further development of tundish technologies to ensure consistency of composition, temperature and flow of liquid steel
- Tundish technologies to achieve steel cleanness (non-metallic inclusions)
- New developments for submerged entry nozzles
- Ecology and production of refractory materials: RHI

The leading producer of high-quality ceramic refractory products is the concern RHI /refractory group Veitsch- Radex- Didier which has customers in 150 countries throughout the world. RHI's products come from 60 companies on 4 continents and are intended for use in high-temperature processes (above 1200°C) in the production of steel, glass, cement, aluminium and non-ferrous metals, and in the petrochemical industry. With their multi-brand strategy RHI offer their customers the complete range of products required in solving individual problems of fire resistance on the basis of raw magnesite, sintered and melted materials, lining and repair masses, mortars, burned articles with chemical and carbon bonding, and prefabricated and functional articles on the basis of magnesite, dolomite, high-alumina, zircon. An active policy of "sustainable development" and constant raising of intra-firm standards on environmental protection are an integral part of the company profile, which has the following focuses:

- delivery of high quality products with perfect thermal properties, hence contributing to the improvement of customers' ecological standards
- constant renovation and modernisation of equipment for the production of refractory materials, as well as steps to reduce emissions and to observe the limits prescribed by legislation, which guarantee the protection of water, ground and air resources. Measures aimed at rapid recovery of land previously used for the open mining of raw materials would be an example of this.
- evaluation of economic and ecological factors prior to a company's decision to consolidate their position on the world market; assessment of the economic advantages involved in the effective use of energy and rational consumption of resources.
- reduction of all forms of contamination of the environment to levels lower than they are at present.
- rules, saving of resources, effective recycling of production waste as the ultimate goal of company strategy.

The 'council for ecology' is responsible for the description of targets, for co-ordination, and the introduction of corrections to existing standards. This council consists of representatives from senior management, the main ecologist, the authorised ecologists at the individual plants, and representatives from the legal affairs, purchasing, sales and research departments of the concern. When planning suitable measures, the ecology council is guided by European standards, in particular by the guidelines of IPPS (Integrated Pollution Prevention and Control, also known as IVU (Integrierte Vermeidung & Verminderung der Umweltverschmutzung).

The system of ecological monitoring that is developed controls the process flows in the time, fixes and archives measurements according to legal requirements. Reliable measurements data are obtained not only from the relevant state or public sources, but, primarily, from our own units and equipment, installed near the company's plants. In addition, RHI finances various systems of independent monitoring, as for instance, the network of bio-indicators in Austria.

270

Research and development at RHI

The protection of environment begins with product development and selection of the raw-material base. To guarantee ecologically safe production, the following factors have to be monitored at the initial stage of new product design: "ecological compatibility" of materials, taking into account possible risks to the raw material or to the materials involved in the process of production, by an auxiliary /bonding agent. In addition to this, the probability of pollution and wastes is analysed on the basis of the production technology and packing processes employed. Considerable attention is paid to staff suggestions. For example, in the last two years 670 work-improvement suggestions made by the German and Austrian colleagues were put into effect; one third of these were connected with the protection of environment.

Not only does RHI require high ecological standards of their products, but their customers also expect these standards. As a result, all suppliers of raw material are checked with regard to ecological standards, and in particular with regard to dangerous substances. Furthermore, ecological aspects and data are given on the material data sheets, and indications are given on the packaging of any possible dangers the material may present to humans or the environments. The main concern remains the observance of the ecological standards of the country in which the product will be used.

Air protection

During the manufacturing process the refractory materials penetrate the numerous high-temperature processes connected with high heat consumption and the corresponding CO_2 pollution. Each manufacturing plant is equipped to perform a cadastral survey of emissions, which fixes the sources of emissions. Cleaning technology is installed which serves to reduce primary energy consumption, to catch dust in woven filters and to recover the dust into the production process, the post-combustion of organic compounds at temperatures above 700° C, and to catch of suspended particles by means of electric filters.

Water protection

The regional regulations concerning water-use, permission for separate installations and/or aqueous cadastral surveys serve as the legal basis for the use of water. A large proportion of the water required is provided by our own deep bore holes or reservoirs; water consumption is minimised by locked circulation and by the cascade use of industrial/fresh water. The moist biotope "Grissener Moor" with 80 species of rare birds, located directly at the water-cleaning facilities in one of RHI's manufacturing plants, serves as an example of cautious water-use.

Waste utilization

Our main purpose remains the minimisation of production waste and its recovery back into the production network, as regulated for the relevant production technology. Besides the selective utilisation of "secondary raw materials" within the concern, RHI's suppliers have also begun to operate different systems controlling the movement of waste. Production concepts for waste utilisation have been developed and introduced in most of the plants, despite the fact that this is not compulsory for many of them.

Dangerous substances

All materials, including dangerous substances, are entered into the data bank and via Intranet they are then used to check the ecological situation regarding the application of various substances in each plant. Rotation and the transport of dangerous substances are regulated by the legal requirements in the individual countries concerning work protection, chemical influences, ionising radiation and the use of a waste classification system, based on the European catalogue of waste. All dangerous substances in each plant are noted down in the special cadastral survey, the substances are stored in accordance with the relevant safety requirements, are employed only as permitted by the regulations and are monitored by the relevant organisations.

Centre of Work Protection and Ecology

This centre was founded in the Radenthein-plant in order to monitor the reciprocal effects of ecological situation and work sites. The primary task of the centre is to evaluate the harmfulness of substances on the human organism, to record current pollution and to develop measures to reduce or eliminate them. In addition to this, the centre is responsible for the development of ecological standards on the rights of membership in the permanent council for coal mining and other mineral obtaining industries of the EC in Luxemburg, EUROMINES in Brussels or ICOH (International Commission for occupational health) in Singapore, and the adaptation of such standards in the entire technological production network of refractories, including mining and storing of raw material, enrichment of masses, moulding, drying, kilning and treatment.

Ecology and production of refractory: examples

All manufacturers of RHI are certified on the standard ISO 9001.2000. The fact that the plants are spread all over the world, makes it possible to narrow the tasks of production tasks for each plant and to achieve high specialisation by concentrating groups of products at the individual plants, which has had a considerable influence on resource saving. Examples of innovations in environmental protection in a number of plants:

Trieben – unique technology is used to produce "oxychrome-sinter". The observance of high ecological standards is confirmed by independent studies of coniferous trees in nearby forests.

Breitenau – a method of dry adsorption of exhaust gases in combination with contemporary filters and thermal post-combustion was patented and introduced.

Hochfilzen, with an approximate output of 100.000t of magnesite/a - the downhill transport of raw magnesite by cableway made it possible to simultaneously produce electricity which was then fed into the local electricity network. The waste department is involved in the operational recovery of land previously used for mining and in restoring it to its original state as Alpine meadows. All exhaust gases pass through dust collection and desulphurisation facilities (method TIMAG-RCE), which meant that emissions were significantly below the level legally required.

Clydebank /Scotland - the requirements of British Environmental Protection Act-1990 and BS 7750 were strictly adhered to.

Radenthein - a reduction in the dustiness and contamination by sulphur dioxide to a level considerably lower than the limits set for health resorts.

A typical feature of the "ecological portrait" of RHI is that most of the European plants are situated in densely populated areas or leisure regions, which superimposes a special imprint on control and process management not only from an economic, but also from an ecological point of view.

Ecology and use of refractory: RHI

As described above, the potentials for radical improvement in the ecological situation must be sought not in the production, but in the use of refractory materials. The special role of RHI - products is to guarantee best performances in terms of reduction in specific consumption, slag-free tapping, increase of lining life, bath agitation by inert gases, steel purity without non-metallic inclusion, reduction of technological metal wastes caused by time loss due to the replacement of refractory material, and to guarantee this whilst, at the same time, aiming for a reduction in energy consumption and loss, and an improvement in the quality of steel produced.

Specific consumption of refractories

A constant reduction in specific consumption and an increase in the durability of brick linings is directly connected with the improved quality of refractory materials and an automated monitoring of movement and combination of materials, as expressed in the equation "scrap + hot metal + additives –slag + refractory = clean steel". The comparison of the converter technologies employed in various regions (Table 3) with the records of more than 30.000 heats in USA, demonstrates the positive influence of an increased MgO-content in the slag on the specific optimum for increasing brick lining life. On the other hand, the comparison of specific consumption figures (Table 4, see NAFTA, EC and Japan) does not show any major differences. This means that, bearing in mind specific regional technological features, the specific optimum has already been reached and a radical change in the situation in the NIS-countries is possible only with increased investments in

monitoring of the steel making technologies. The situation cannot be changed only by one sintered periclase in the refractory material and by MgO-saturation in the slag of more than 15%. The situation is different in the case of refractory consumption for EAFs (Table 5): in contrast to the case of the approximate 25 BOF-vessels in operation world-wide, the transfer of electric steel making technologies became 'big business' long ago for the approximate1600 EAF-unit in operation today. The most immediate processes today are the reduction of Power On- time and bath agitation (C/O_2-lance, bottom and side wall stirring by inert gases).

Table 3 Regional differences in BOF-technologies

	Europe	North America	Japan
Heats/day	25-35	15-20	21-40
Coating	Seldom	yes + slag splashing	seldom
°C of tapping	1660-1730°C	1650-1680°C	1650-1700°C
Additives	dolomite/magnesite	dolomite	Low % of slag
% of MgO in the slag	2-5	8-14	10-15
Si of hot metal, %	<0,7	<0,8	<0,3
Dephosphorisation	No	no	Yes
% of scrap	~20	~30	~5
Bottom stirring, %	100	60	100
Current lining life	1000-4000	6500-33000	2000-5500

Table 4 Specific refractory consumption in converter steel making by regions

Region	Country	steel, 1000t	specific consumption, kg/t		
			bricks	mixes	total
America	NAFTA	68.400	0,49	0,73	1,22
	South America	24.300	0,73	0,27	1,0
Asia	China	69.000	1,7	1,0	2,7
	India	12.600	3,7	1,0	4,7
	Japan	70.000	1,0	0,4	1,4
	South-East	41.000	0,89	0,4	1,29
Europe	NIS	39.500	2,8	10,5	13,3
	EC	99.800	1,0	0,36	1,36
	Eastern Europe	25.400	2,68	0,7	3,38
Africa	South Africa	5.600	1,46	0,47	1,93
Total, 1998:		**465.180**	**1,36**	**1,42**	**2,78**

Table 5 Specific refractory consumption in electric steel making by regions

Region	Country	Steel, 1000t	specific consumption, kg/t		
			bricks	mixes	total
America	Central America	430	2,7	9,6	12,3
	NAFTA	56.000	1,49	4,48	5,97
	South America	12.240	1,55	6,49	8,05
Asia	China	18.000	2,0	4,5	6,5
	India	8.000	2,8	5,5	8,3
	Japan	34.000	1,0	3,5	4,5
	South-East	9.200	1,8	3,5	5,3
Europe	NIS	8.110	5,34	11,95	17,28
	EC	53.100	1,11	4,16	5,27
	Eastern Europe	7.350	2,11	6,67	8,78
Africa	North Africa	2.820	3,10	7,83	10,93
	South Africa	4.200	1,8	6,5	8,3
Total, 1998:		**254.180**	**1,6**	**4,8**	**6,4**

Inert gas stirring

Bath agitation by inert gases is a crucial part of contemporary metallurgical processes in view of the significant influence on all subsequent technological operations and on steel quality.

Bottom stirring in BOF-vessels

Taking into account the fact that the almost all converters in Western Europe are equipped with bottom stirring, the potential to improve ecological standards in the steel industry of NIS countries becomes obvious. The technologically substantiated amounts of specific consumption of inert gases for the bottom stirring in the BOF- route vary from 0,03 to 0,08 $Nm^3/min*t$. In contrast to the existing practice of stirring by plugs with one channel of gas supply, the use of Multi-Hole-plugs with a number of tubes till 32 pc. is predominant guaranteeing a reliable supply till 160 m3/h.

Direct bottom stirring in EAF

The Direct Purging Plug (DPP)-system (Fig. 1) was employed for the first time in an EAF at the beginning of the 80s at the plant in Beltrami/Italy [6] and designated one of the landmarks of the modernisation of electric steel making reactions, involving active bath agitation by inert gas leading to specific consumption reduction (Table 6) due to the rapid homogenisation of the bath as regards temperature and chemical composition and intensification of the slag/metal.

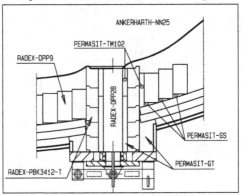

Fig. 1. DPP-system for direct bottom stirring

Table 6. Reduction of specific energy consumption in an EAF by DPP-systems

EAF	Steel grade	Before	Save with DPP
160 t	C-, Cr-Ni	630	30 kWh/t
130 t	C-Steel	550	20 kWh/t
115 t	C-Steel	460	30 kWh/t
100 t	C-, Cr-Ni	575	47 kWh/t
80 t	C-Steel	424,4	27,7 kWh/t
75 t	C-Steel	462	74,9 kWh/t
60 t	C-, low alloy	670	110 kWh/t
30 t	Cr-Ni	550	41 kWh/t

The injection of the inert gas bubbles accelerates reduction on specific pressure of the reactions of decarburisation /deoxidation, as well as the kinetic and thermodynamic reaction of the removal of nitrogen and hydrogen. Since the kinetics of the ratio FeO/[O] depend on diffusion processes on the phase surface "slag/steel", the effect of DPP is expressed as follows: the phase surfaces increase by mixing, and oxygen dissolved in the bath reacts more rapidly with carbon. According to the distribution law, the metallurgical necessary oxygen from (FeO) in the slag is reduced. The free Fe goes over into steel, which increases the yield and decreases FeO in the slag. The typical slag/steel - reactions to remove silicon are determined, first of all, by diffusion processes on the boundaries of phases.

The flow of these processes is the slowest and determines the rate of the whole reaction. The vertical effect of mixing increases the surface of reaction and decreases the diffusion layer between slag and steel, which accelerates the process of the silicon removal. SiO_2 as acid component has negative influence on the fire resistance of the brick linings. The lime (CaO), necessary for the binding of (SiO_2), becomes more rapidly reactive because of the agitation effect, and this results in the additional effect of a higher brick lining life. The catalysation of the dephosphorisation reaction is similar. To obtain de- and tricalcium phosphate it is necessary to have the reaction [P] - (FeO). Since this reaction depends on (FeO), the kinetics of dephosphorisation is improved by vertical mixing. The reaction of desulphurisation has a very strong influence on both high (CaO) and low activity (FeO), which displaces the equilibrium to the side (CaS). Through the better dissolution of lime and catalysation of reactions connected with (FeO), the DPP-system improves the reaction of desulphurisation. The deoxidation of melting usually occurs before the conditioning of slag and the tapping into the furnace or into the ladle. The [O], dissolved in the bath, in connection with (FeO) in the residual slag is connected by additives of Al, Si, SiMn to (Al2O₃, SiO₂, MnO). A high (FeO)/[O] content causes the unnecessary burning out of desoxidants.

Based on Nernst's law of distribution and reaction of decarburisation, the DPP- systems contribute to a reduction of (FeO)-content, and also of [O], and, thus, reduce the consumption of desoxidants, which, depending on their type, quantities, presence and costs can substantially influence the economy of the melting process. At the same time, the reduction in consumption of Al or Si has a positive effect on lining life. Similar effects are achieved by the process of burning out of additives and improvement in their dissolution in the bath. The removal of hydrogen (and of nitrogen) is related to a reduction in specific pressure, to temperature and to the presence of other elements. The solubility of hydrogen increases with temperature and increased TI, V, Nb, Cr - content and falls with increased C, B, Si, P, Al. Nitrogen increases with higher V, Nb, Cr, Ta, Mn, Mo-content and decreases with increased C and Si. The DPP-process optimises these processes: bath agitation helps to diffuse hydrogen and nitrogen into the rising gas bubbles. Even during the use of nitrogen as the carrier gas, the specific pressure of nitrogen by formation of N/CO bubbles is reduced and contributes to the removal of nitrogen. If it is necessary to achieve lower values of nitrogen, argon is used.

All these theoretical calculations find confirmation in practice: the 65t-EBT-EAF with the power 60 MVA (40 MVA in the work) with an oxygen-carbon manipulator and gas-oxygen burner in the door, was built in1993 in Podbrezove and modernised in 1999 by the DPP-system with 3 tuyeres DPP28 with a 3- linear gas regulation station Dose- Module with the Simatic S7 panel. As a result of the installation of the DPP system, the specific consumption of electric power was reduced by 15 kWh/t (476 to 461), and of anthracite to 24 kg/hts (218 to 194), respectively on 0,41 kg/t (3,84-3,43). Power On time was reduced by 1,5 minutes (40,6-39,1), melting time by 6 minutes (62,6-56,5). The yield increased by 0,61% (93,79-93,18). Productivity rose by 5,6t/h (54,3-60). Gas and oxygen consumption remained at their previous levels. Although oxygen consumption remained the same, the process of decarburisation accelerated from 0,173 %C/h to 0,353 %C/h and from 29 ppm C/min to 58 ppm C/min., respectively. Increase in productivity and a lower consumption of electric power contributed to a reduction of 0,25 kg/t (2,69 to 2,44) in the specific consumption of electrodes. The DPP-system consumed argon and nitrogen at a rate of 30-60 l/min, with the specific wear of purging plugs at 0,5 mm/h and by their replacement after 840 heats.

The total costs (refractory materials + Ar/N₂) vary, depending on furnace output (5 - 200 t), between 0,3 and 1 $/t of LS, but save (electric power, time, metals) between 0,8 to 8 $/t of LS (note: there is no linear dependence of the economic effect on the size of the furnace). This system has advantages even for small furnaces which consume (electric/chemical) more than 430 kWh/t LS on the one hand, or, on the other hand, for small 5-20t -furnaces (for example, Ebroacero, Sande Stahlguß, ABB Zamech), in spite of the seemingly high initial capital investments for the bottom stirring system. This is true, particularly, of stainless and special steel production. There is one other factor which is even more important than the rapid return on investments over a period of 3-6 months. This is the fact that, even with high efficiency of 45% for the current power stations and minimum reduction of specific energy consumption due to the DPP-systems it is easy to calculate to what extent CO₂ pollution could be reduced. Other advantages of bath agitation by DPP- systems are:

- reduction of the Cr₂O₃-content in the slag from 9 to 2.25% in 70t-EAF for stainless steel
- 6-8% increase of yield of (t/h) for furnaces 80-100t because of the reduction of FeO in the slag

- 20% less lime consumption for 160t EAF
- 30% longer side wall lining life
- 25% shortening of the tap-to-tap time and Power On-time
- reduction in the consumption of ferromanganese (to 1 kg/t), of ferrosilicon (0,3 to 1 kg/t), O_2
- 25-35% reduction of the content of sulphur and phosphorus, decrease in N_2.

All these points confirm the contribution of electric steel making to "ecological cleanliness". More than 160 EAFs world-wide are equipped with the systems of bottom stirring, of which approximately 95% are the DPP-systems produced by RHI. Most of the projects were realized in the middle of the 90s: between 1994 and 1998, 80 bottom purging systems were installed, 75 of which came from RHI.

Indirect bottom stirring (system VVS) and stirring in open hearth furnaces

The covered (indirect) bottom stirring (system VVS) is an alternative to the DPP-systems. The gas flow in such a system occurs through the porous hearth (Fig. 2). The advantages of covered bottom stirring are that it increases and improves the situation in the following fields:

- melting of the heavy scrap
- melting of the alloying additives
- meeting analysis and temperature of the melting
- C/O-ratio, closer to the equilibrium
- 0,5% higher yield (calculated on a base of 100 kg of slag /t LS)
- higher yield of alloy additives
- desulphurisation / dephosphorisation
- foaminess of slag

Furthermore, this process helps reduce:

- electrode consumption
- melting time (to 3 minutes)
- FeO-content in the slag - to 5%
- specific energy consumption, to 20 kWh/t
- differences in temperatures between the measurements in the furnace before tapping and in the ladle after tapping from the furnace
- retardation of boiling
- consumption of materials for the hearth in the zone of bath agitation
- N-content and tapping temperature

Experience with the VVS-systems showed that there are obvious benefits to be had in terms of reduction of specific energy and refractory material consumption (Table 7). Such systems were installed mainly in EAFs in Japan.

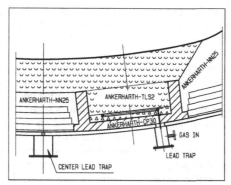

Fig. 2. VVS-system for covered bottom stirring

Table 7. Application of VVS-systems in electric arc furnaces

Item \ capacity of EAF, t	50	60	80	85	90	130	150	175
Tap-to-tap reduction, min	4	10						2
< specific energy, kWh/t	20	25	25					15
< specific electrode consumption, kg/t	0,2	0,2				0,1		0,2
< of FeO in the slag, %					7	5	1	
> yield of alloys , %	3Mn	1Cr	1Mn					
> yield, %	1					1		
< specific refractory consumption, hearth ,%	40	30	50					40
< N -content, ppm						6	27	25
< tapping temperature			35		30	20		8
desulphurisation /dephosphorisation	P				P	S		

The proportion of oxygen-converter steel and electric steel in the total volume of steel making in Russia in 1999 was only 72,1% [2] of the previous year's figures. This was mainly due to the reanimation of capacities of the open-hearth furnaces. In view of the positive experience gained in applying VVS-bottom stirring systems in EAFs (Table 7), the firm TECHCOM/Germany, together with RHI, developed technology for covered bottom stirring of open-hearth furnaces.

The magnesite fritting of hearths and, in certain cases, the brickwork are destroyed in the classical open-hearth process, as becomes particularly apparent after stoppage for cold repair [7]. The individual fritting sections are converted into fine dispersed powder. Because of the serious damage to the hearth it is necessary prematurely to stop the campaign for hot/cold repair of hearth, which leads to an increased consumption of magnesite powder (up to 11 kg/t) and to time losses (2 % of calendar time). The open hearth technologies with less hot metal or transfer to the carburettor process (which is actual in the contemporary economic situation) hampered the process of the dissolution of slag additives, and difficulties appeared in the heating of metal (including temperature contrast in the volume of bath), the effectiveness of desulphurisation decreased, and metal contamination by non-metallic inclusions rose. All these problems were reason enough for the development of new technology which enabled bath agitation by the inert gases in the open-hearth furnace through the porous hearth. TECHCOM proposed, in combination with the purging process, the introduction of new technology which involves laying refractory material, based on the Ankerharth- mixes. The major advantages of such mixes lie in the natural qualities of the source material: in its petrographical special features, its granularity, and the combination of the low SiO_2-content with a large quantity of Fe_2O_3 and CaO.It is a well known fact that the Martin method of melting is based on the technology of mixing the layers of metal in the furnace. The rate of the process is determined by the speed of its slowest kinetic part - in essence, the chem reactions through slag, and temperature distribution from the burning gases into the fluid bath through the slag by convection.

During mixing by purging the gases, the kinetics of all "metal –slag" reactions involving (CaO), [P], [S], (FeS), (Fe$_2$O$_3$), [O], [C] and the elements present in the scrap is strengthened. Agitation by gas contributes to the homogenisation of liquid steel and, creating on the possibility a large quantity of small bubbles and thus producing the effect of flotation, is simultaneously rendered the refining action. The speed of steel decarbonisation rises, cause reaction is initiated earlier. The melting temperature is equalized throughout entire volume. Because of the homogenising action of bottom stirring not only kinematical but also thermodynamic effects are achieved. The differences in element concentration caused in the process of diffusion are removed and approach metallurgical equilibrium. Mixing also contributes to the derivation of non-metallic inclusions into the slag. As a result of bottom stirring, the melting process is shortened by 20 minutes. Depending on organisational measures, the composition of the metal-charge, decrease of times losses of equipment, there are still opportunities for improving this index. The average weight of melting increases by 2 t. Fuel consumption (excluding the line of low pressure and correction factor) decreases to 30 $/t of steel.

The duration of hot repairs is reduced by 0.2 % (calendar time). The specific consumption of refractory powders decreases to 5 kg/t, including new material (0,8 kg/t) and servicing materials (4,2 kg/t of steel). The given indices do not demonstrate all improvements on furnace productivity, flexibility of the system under different conditions, extended lifespan of refractory masses or optimisation of the operating temperature of the refractory materials in the main arch of the furnace.

After industrial trials with bottom stirring using the VVS system, the plant in Liepajas, the Russian steel works Nizhny Sergy, Revda, Izhevsk. Asha, Chusovoy, Taganrog, Vyksa, and Ukrainian plants in Dnepropetrovsk and Donetsk modernized their open-hearth furnaces by bottom stirring. The process of Direct Side Stirring for open-hearth furnaces is still in its trial period. The first test phase has been completed successfully in Liepajas, and the technological finishing of the operational characteristics of method is conducted.

Flow Control

Together with the above systems for the primary metallurgical aggregates, the value of the time factor of tapping for the quality of the product rises in proportion to the "forthcoming" of steel to mould. The time factor involved in the replacement of damaged submerged nozzles and metering nozzles for opened pouring is expressed by a reduction in the quality of the steel as a result of reoxydation, an increase in the non-metallic inclusions, a reduction in yield due to rejection of the slab sections where submerged nozzles /shrouds have been replaced. These disadvantages are solved today, to a large degree, by the installation of the "quick change" systems for flow control: such systems radically reduce the factor of refractory material in the ingot quality. The system for submerged nozzles quick change, Interstop's SNC-N, effect replacement of the submerged nozzles in less than one second. This is made possible by a special design of the refractory elements, so-called Tight Push Edges.

The flexibility of production and high productivity with a minimum of costs is limited in many respects by the duration of the sequences, caused by clogging and wear of the nozzles. Systems MNC solves such problems by on-line replacement of the worn articles; moreover - because of the "off-line" maintenance of the mechanics only a minimum number of complete sets are required to ensure operation of a CC-machine. As a rule, the quick change systems are mounted on the basic centring plate from a cart and fixed by unwedging or by bolt assemblies. Depending on the requirements of customer, the design of the system provides for use in two different types of refractory lining, either with the well block installation on the inside of the tundish (standard) or with the well block installation on the outside, so-called "hot replacement". The compound-blocks with the co-pressed zirconium inserts are stretched by metallic cowling.

If necessary, the MNC-systems can easily be modified to the system MNC- SN where the steel flow is protected by installation of an SN lock. Moreover, the replacement of the metering nozzles is produced in the "pouring position" of submerged nozzles. By releasing the blocking lock SN, it is possible to replace the pair "metering nozzle – submerged nozzle" by one movement of the hydraulic cylinder. Record-holders in the application of such quick change systems are SAM Montereau/ France - 45 h (53 hts), Lechstahlwerke/ Germany -48 h, BHP Sydney/ Australia -90 h (79 hts), Gerdau Barao de Cocais/ Brazil -50 h.

For CC-machines equipped with slide gate flow control, is possible to combine slide gate operation and quick change of submerged nozzle using Interstop's system 33QC- SNC. This solution combines the above mentioned advantages of the SNC systems with the benefits of the slide gate casting, for instance, the fact that there are no problems with non-closing of the monoblock stopper. The sliding line of the middle slide gate plate and the line of replacement of the submerged nozzle are perpendicular to each other, the independent hydraulic drive of the slide gate cylinder and of the SNC-system made it possible to replace a submerged nozzle in no more than 2 seconds, without interrupting the casting flow or lifting of the tundish. Casting by one complete set of plates reaches 30 hours or more.

Conclusions: trends on the economy and ecology of refractory material

The quick change system for flow control, and the bottom stirring systems described earlier, require the application of high-quality refractory materials. In particular, articles for flow control must ensure high-precision contact between exchangeable submerged nozzle and the basis tundish nozzle along the whole sliding surface (8) as well as having high wear resistance.

Of course, such materials are more expensive, and in each case it is necessary to evaluate the price/quality ratio before applying the "expensive" refractory materials. However, the latest information on 1,8 billion € assignments from the EC and the European Bank of Reconstruction and Development to be invested in the ecological regeneration of northwest Russia [9] involuntarily raises the question: "Is it not better to invest money in the modernization of production and "expensive" materials before the ecological situation of the region becomes critical, than "afterwards, with fortitude" to spend the same money on overcoming the ecological problems which have emerged ?" RHI's main concern in this connection is to choose selectively the materials needed for the complex solution to fire resistance problems for the specific situation, at the same time maintaining a leading position in the production of high quality refractory materials. Furthermore, the product must remain competitive whilst strictly observing regional and international ecological standards.

Between the end of 2002 and the end of 2003, the European plants concerned will have received certification to ISO 14001 standard, confirming that they have achieved these goals. Company-internal standards are aimed not simply at observing existing standards, but at achieving additional success and improvements in the field of environmental protection and in the ecological harmony of living environment as a place of work and living space.

280

Notes

[1]http://prometal.com.ua/analit.php3?n=28&p=04 Pronin, E. Ecology in the steel industry: time to utilise of non-utilizing waste.

[2]http://www.mnr.gov.ru/text/4/Gosdoklad99/show_doc.php?gid=4&file=Part5-1.htm   Part V Influence on environment by industries

[3] Klimaschutzerklärung der deutschen Industrie unter neuen Rahmenbedingungen, Monitoring-Bericht-99, Fortschrittsberichte der Verbände, RWI Essen

[4] Hillebrand, B., Buttermann, H.G., and Oberheitmann, A. (1997 First Monitoring Report: CO_2-Emissions in German Industry 1995-1996, RWI-Papiere, N50, ISSN 1433-9382, Nov.97 Essen

[5] Technology Road Map to Determine the Research Priorities of the European Steel Industry, Chapter 3.2; 1999 EUROFER Brussels http://www.eurofer.be

[6] Faggionato, A., Galenda, A., Mario, F., Pawliska, V. and Cappelli,G. (1990) 1 1/2 Jahre DPP-Spüelen im 125t EBT-Ofen bei Beltrame. Sonderdruck aus „Radex-Rundschau", Heft 1/1990),

[7] UDK669.183.2:669.046.5 Kupshis, E.: Bottom stirring of open-hearth furnaces by VVS-system

[8] Smirnov A.N., Glaskov A.Ya., Pilyushenko V.L., Efimov V.A., Brodsky S.S., and Pikus M.I. (2000) Theory and practise of continuous casting. Donetsk: DonGTU, OOO»Lebed» 2000, ISBN 966-508-244-2, p.199.

[9] «Ecology in Russia gets 1,8 billion €», «Dengi»/Kommersant; №27(382)17.07.2002

CHAPTER 27
Usage of mining wastes for open pits' reclamation

Vadim Zoteev, Tatyana Kosterova, Alexandr Medvedev

Department of Hydrotechnical Facilities,
Russian R&D Institute for Complex Utilization and Preservation of Water Resources
23, Mira St., Ekaterinburg 620049
Russia

Summary

This chapter offers some suggestion on the use of mining wastes for the reclamation of open pits based on some case studies.

Introduction

Information and methods on how to reclaim technogenous landscapes, particularly those occupied by developed open pits, is very important for large mining regions, amongst them the Ural region of Russia.

The most effective way to reclaim pits, situated on the sites of working ore mines or processing plants, is to fill them in with mining wastes, in particular, tailings of ores' enrichment. In such cases it is possible to refuse certain substances from the construction of special surface tailings storage facilities which are dangerous and sources of environmental pollution.

The first major project in the Urals reclaiming land disturbed by mining was carried out in Nizhniy Tagil City, Sverdlovsk oblast'. By the middle of the 90s a difficult ecological situation had arisen in the city due to the severe and long-term environmental impact of large mining and metallurgical plants.

The most large-scale geological environmental changes in the city were connected with mineral resources quarrying. Within the city limits there were 6 developed deposits of ferrous and copper-pyrite ores. At present, mining continues for ferrous ores, limestone and building materials. The land occupied by mining wastes within the city boundaries comprises 30% of the total area of the city. The height of waste piles, which have remained after development of the deposits, is as high as 50 m. The depth of developed open pits ranges from 50 to 280 m. These objects are sources of atmospheric air pollution, soil, surface and underground water contamination. The total area in the city occupied by different wastes is about 3000 hectares.

One of the major sources of environmental pollution in the city was the Visokogorsky ore mining and processing enterprise (OMPE). Several completely developed open pits were a part of OMPE, including Main Pit, the volume of which was 70 mln.m^3.

The Cheremshanka tailings storage facility was also situated on the OMPE site. Here, 40 million tons of ferrous ores enrichment wastes from the Vysokogorskoe deposit were stored. These ferrous ores in addition to iron, also contained small quantities of copper, which could not be extracted

281

W. Leal Filho and I. Butorina (eds.),
Approaches to Handling Environmental Problems in the Mining and Metallurgical Regions, 281–288.
© 2003 *Kluwer Academic Publishers. Printed in the Netherlands.*

because the necessary technology was not available. For this reason, the ferrous ores enrichment tailings contained up to 0.35% of copper (average 0.25%).

When the Cheremshanka tailings storage facility was almost full, it was extended to cover an area of 239,7 hectares. The main part of the tailings was in a dry state. As a result, the tailings storage facility was a strong source of atmospheric air pollution in Cheremshanka village, which was situated nearby.

There were no doubts that both the pit and the tailings storage facility were in urgent need of reclamation. However, some serious investments were necessary before this could be implemented.

Another problem was that, if the Cheremshanka tailings storage facility was to be closed, it would be necessary to build a new facility for the storage of ferrous ores' enrichment tailings because OMPE was still mining at the mine "Magnetitovaya", located below Main Pit.

In order to find a solution for this situation, a project aimed at an economic-efficient reduction of man-caused impact on the environment at the plant site was proposed. The basic principle of the project was that it was to be self-financing.

It was suggested that 3 tasks be solved concurrently:

1. Useful components extraction (in this case – copper) from tailings stored in Cheremshanka tailings storage facility and from newly manufactured OMPE tailings into commodity output.
2. Ecologically non-hazardous secondary wastes, formed during copper extraction, to be placed in the worked out space of Main Pit.
3. Reclamation of the tailings storage facility and the pit financed by the money received from the sale of copper concentrate.

The chance to fulfil the task successfully was given primarily by the existence of the flotation department at OMPE, which was able to process up to 800 thousand tons a year of ore raw material containing copper concentrate.

The large volumes of copper-bearing tailings stored in the Cheremshanka tailings storage facility, together with newly manufactured tailings from ferrous ores enrichment, could provide the flotation department with essential loading and profitable copper concentrate production.

The main pit had the necessary space for the fulfilment of the project. Besides, the pit was situated at only a short distance from the flotation department and it was possible to direct secondary wastes there by self-flowing, which considerably reduced exploitation costs.

Thus, in principle, it was possible to implement the project. However, before it could actually be put into practice, it was necessary to solve several problems of a technological and ecological nature.

Flotation was selected as the main technological operation for the reprocessing of mining wastes. As the copper content in original raw materials was low, it was necessary to improve the technological flotation process in order to obtain an acceptable level of copper extraction into commodity products. In this process, the flotation reagents' consumption must be such that their content in flotation outflow, which was to be directed into Main Pit, would not exceed established standards.

As a result of the investigations carried out, a method of poor copper-bearing wastes flotation was created with the receiving concentrate containing 25% copper, and with a level of copper extraction of about 65-70%.

As the content of sulphur in wastes did not exceed 1.5%, and copper did not exceed 0.35%, it was recommended that 2-ethylhexanol be used as a frother, and butyl xanthogenate of potassium as a collector. Simultaneous usage of these reagents made it possible to significantly reduce consumption, which increased the ecological reliability of the technology and provided a considerable reduction in the concentration of residual reagents in the flotation's outflow.

Additionally, water-circulating systems were used, the flotation outflow's cleanout used activated carbon (which was removed from the process and later added to the copper concentrate) and their next dilution in 1:25 ratio by non-sulphur tailings from the ferrous ores magnetic separation process. All these measures made it possible to ensure that the content of flotation reagents 'in the tailings directed into Main Pit was considerably lower than the established standard level.

The following factors complicated the placing of flotation tailings' in Main Pit:

1. The presence of karst limestone at the pit edges and bottom, with a filtration coefficient up to 50 m/day and higher.

2. The emergence of a caving zone at one of the pit edges, which was formed during mining of "Magnetitovaya", situated below the pit.

3. The existence of a direct hydraulic connection between the workings of the mine "Magnetitovaya" and Main Pit, due to which all atmospheric fall-outs in the pit area were collected in the water drainage of the underground mine.

4. The execution of mining works at the "Magnetitovaya" mine at a depth of 380-540 m below the bottom of Main Pit.

Due to the above-mentioned reasons for pit-mine complex preparation for pulp receiving, mining-technical and hydro-geological investigations were conducted, on the basis of which the following engineering and technical measures were developed and carried out.

a) Concrete lintels were designed and constructed in the underground workings connecting "Magnetitovaya" with the workings of Main Pit's drainage mine. This meant it was possible to exclude the possibility of pulp breaking into underground mining areas.
b) A new pumping system was built and Magnetitovaya's pumping system was reconstructed in order to ensure that the maximum calculated volume of filtrated water could be pumped out.
c) Geo-mechanical calculations were made to evaluate Main Pit's edge massive filtrating characteristics reduction during the process whereby it is filled with the tailings.
d) Design and construction technology of between-layer and edge screens made of clay soils was developed, making it possible to regulate water losses towards filtration within given limits.
e) A closed water pumping system and circulating water supply, which included the flotation department, the pit and "Magnetitovaya", were designed and built.

To ensure secure and ecologically clean tailings' stocking in the pit, a schedule of the filling which took place was introduced. This included the following requirements.

1. In case of filtration losses in excess of the acceptable limit (1600 m^3/h), it is necessary either to reduce the water column height above the tailings or to temporarily stop pulp drain for anti-filtration screen formation.

2. Anti-filtration between-layer screens are to be formed from the clay pulp discharge into the pit, which should be of sufficient volume to provide dust-clay material layers no less than 0.3 m thick, along the entire surface of the tailings.

3. To prevent water leakage through the caving zone at the pit's edge above the +72 m mark, it is necessary to construct an inclined anti-filtration screen there, made of compacted clay soil and no less than 2 m thick. To increase the screen's stability, compacted clay soil must be loaded with rock. Screen build-up is to be made in waiting lines of maximum height 25 m.

4. After the pit has been filled in to the desired level, the surface of the tailings must be covered with an anti-filtration clay soil screen that prevents the penetration of atmospheric fall-outs into the waste.

At the beginning of 1995 there was an organised experimental-industrial test of the project's activities which showed that the proposed copper extraction from mining wastes was ecologically safe, and tested the results of the pit water filtration processes into the underground mine works using a model.

On the basis of the results of this investigation, possible water loss from the pit – in the case of an increase in its filling level - was calculated. The data showed that losses would not exceed:

- $700 \ m^3/h$ if the water column height above the tailings is no higher than 10 m;
- $900-1000 \ m^3/h$ if the water column height is 12 m;
- $1300-1500 \ m^3/h$ if the water column height is higher than 15 m.

It was also shown that, in the case of a between-layer screen inwash, filtration loss is reduced to $300-500 \ m^3/h$ after the tailings' surface reaches a mark +10 m, and that this gradually increases in relation to the rise in the water level

Results obtained from observation of filtration loss between May 1995 and February 1998 fully proved these forecasts.

In July 1995 this complex system involving ecologically safe mining wastes reprocessing technology and stocking of flotation tailings was accepted in Main Pit into continuous exploitation. During the period up to 2000, the OMPE flotation department reprocessed 2,600 thousand tons of mining wastes and received 18,893 tons of copper concentrate, with a copper content of 26%. During the same period 9,100 thousand tons of mining wastes were stocked at Main Pit, including non-sulphur tailings from the ferrous ores magnetic separation process.

Filtration loss at Main Pit was regulated with the assistance of a calculated water column height support above the tailings and a transportable pumping system. During the period in question, filtration loss and reagent content in mine water never exceeded the calculated quantities.

Therefore, the realization of this project involving the Cheremshanka copper-bearing tailings storage facility, tailings reprocessing and stocking of secondary wastes at Main Pit, all tasks were successfully completed:

- Work at OMPE was able to continue without a new surface tailings storage facility being built;
- The Cheremshanka tailings storage facility was almost fully reclaimed, and the technical stage of Main Pit's reclamation was begun;
- The ecological situation in Nizhniy Tagil improved significantly.

Figures 1 and 2 show schemes and photographs of the Cheremshanka tailings storage facility and the condition of OMPE Main Pit in 1995 and 2000.

a) before the beginning of reclamation (1996)   b) after reclamation (31/01/2000)

Fig. 1. The reclamation of the Cheremshanka tails storage facility (Visokogorsky ore mining and processing enterprise, Russia, Sverdlovsk Oblast, Nizhniy Tagil City)

It is essential to point out that, despite the ecological orientation of the project, it was not initially accepted by the population of Nizhniy Tagil. The project was of considerable social interest and there were a number of discussions on the necessity for and advisability of carrying out such a plan. In particular, representatives of the green movement were opposed to the reprocessing of mining wastes at OMPE's flotation department, fearing potential environment pollution with flotoreagents. They were also opposed to the idea of filling Main Pit with secondary wastes. An alternative suggestion was that a mining museum be set up on the pit site. Reasonable fears of possible underground and regional surface water pollution due to the filling in of Main Pit flotation tailings were also expressed.

For these reasons great care was taken to study public opinion during the project's environmental impact assessment procedure. The main purpose of the investigations was to conduct a "quality" survey, followed by a "quantity" assessment.

For the "quality survey" groups of people were formed who were professionally connected or acquainted in their daily lives with the above mentioned problems. These included representatives of the city administration, plants managers, environment and sanitary supervisors, members of the city's "green" movement.

A discussion about possible ways of solving the ecological and social problems in Nizhniy Tagil and OMPE by means of the proposed mining wastes' reprocessing project was organised with these groups of people.

The "quantity assessment" consisted of a questionnaire to be filled in by respondents representative of the Nizhniy Tagil population, followed by evaluation of the replies. 500 respondents were selected randomly on the basis of statistical chance, using address lists. Calculation showed that the rate of error involved in such a selection procedure was no more than 3.4%.

On the whole, the results of the surveys showed that a decision in favour of the project was rated in the category "unpopular" by the respondents. The main reason for this was the low level of awareness concerning the content of the project, and positive and negative aspects of the proposed solutions, in both economic and ecological respects.

In view of this situation, the administration of OMPE, together with the authors of the project and some specialists undertook popularization work through the Nizhniy Tagil mass media. Several conferences involving all interested parties were organised to look into the ecological and social-economic problems of implementing the project.

This finally made it possible to reach agreement on the question of the ecological and social-economic acceptability of the suggested project, and to make a decision in favour of implementing it.

The success of this project became a basis for its implementation at other mining plants in Russia. In particular, the OMPE administration in Gaiskiy (Gay city of Orenburgskaya oblast') decided to reject the idea of constructing a new surface tailings storage facility construction, and in favour of locating the ore enrichment tailings in a worked out space of pit № 2 and of the subsequent reclamation of the pit and the old OMPE tailings storage facility.

During the project's environmental impact assessment procedure in 2000 a suggestion was made concerning the possibility of ecologically safe tailings stocking in the pit on condition that certain project corrections were made in order to reach the aims listed below:

1. Provision of long-term functioning of the mineral water health resort "Gai" situated near pit № 2 by preserving the water's chemical composition;
2. Prevention of underground water pollution and possibility of underflooding neighbouring areas;
3. Long-term preservation of the mineral and chemical composition of the tailings stocked in the pit, which are prospective raw material for the extraction of copper and certain other mineral components which they contain;
4. Development of the tailings storage facility conservation method, ensuring minimal costs for its future reclamation and, if necessary, involvement in reprocessing.

If the "Gai" health resort is to function in the long term, it will be necessary to ensure not only that the chemical composition of the water is preserved, but also to guarantee the preservation of water resources preservation, which have been greatly depleted due to the development of a vast depressive funnel in the vicinity of pit № 2.

During the process of filling the pit with tailings, the size of the depressive funnel will inevitably decrease because of filtration wastage from the pit. Investigations showed that, with a rise in the water level in the pit above the +330 m mark, the presence at the pit edges of permeable Jurassic sand layers would render intensive filtration of the tailings liquid phase towards the health resort "Gai" inevitable. This would lead to a desalination of the medicinal waters and to a deterioration of their quality.

If the level of the pit in-filling is limited to +330 m, the restoration of the natural water level in the area of the health resort will be ensured by technogenous (under waste pile) water infiltration, the chemical composition of this water being identical to natural medicinal waters. So, the composition and medicinal waters' resources will be retained.

Should pit filling above the +360 m mark occur, the sections of the pit edge which exceed the +330 m mark must be closed off by an anti-filtration screen, in order to protect the health resort from underflooding with filtrate from the tailings storage facility. In this case, however, it would be impossible to guarantee the long-term quality preservation of the medicinal waters, although the process of their degeneration can be significantly slowed down.

The danger of the region's underground waters being polluted and the surrounding area being underflooded can safely be counteracted by limiting the pit's filling level to the +330 m mark. This level is below the local off-loading basis (the Kolpachka river shore line mark is +359 m), which excludes the migration of pollutants and the underflooding of rock mass outside the boundaries of the existing depressive funnel. On condition that certain requirements concerning the tailings' inwash technology and waterproofing and drainage creation be fulfilled, the safe level of filling can be increased up to the + 340-345 m marks.

According to the results of the tailings chemical analysis, the average copper and zinc content in the tailings is 0,29%, which is somewhat higher than in the Cheremshanka tailings storage facility, which is already involved in reprocessing with the goal of copper concentrate production.

Because of the scarcity of copper-zinc raw materials slag dumps in Kirovograd and Sredneuralsk copper-smelting plants are involved in secondary processing. It is very probable that, in 10-15 years at the latest, the question of tailings reprocessing at Gaiskiy OMPE will be raised. In view of this possibility it is necessary to envisage a method of tailings' conservation by which they will be safely protected from leaching by atmospheric fall-outs. This will ensure the long-term preservation of the mineral and chemical composition of the tailings and provide an opportunity to reprocess them profitably and to extract useful components from them.

In order to fulfil this task, it will be necessary, after the pit has been filled up to the +330 m mark to close off the tailings with an anti-filtration screen made of clay soil to prevent or significantly reduce the penetration of atmospheric fall-outs. This will prevent sulphide minerals from oxidation and transition into sulphates.

Then there will be two possible alternatives if secondary tailings processing is necessary:

1. If the reprocessing is started within 8-10 years after conservation then the main method of copper extraction will, most likely, be flotation. In case technology is employed similar to that used in Visokogorsky OMPE, it will be possible to provide a high level of copper extraction from tailings and a low flotoreagents' content in secondary wastes. A hydraulic dredge can be used for the exploitation of tailings in the pit.

2. If secondary development is begun only after 15-20 years or if an anti-filtration screen is not installed during the process of tailings storage facility conservation, then, due to the high degree of sulphides oxidation, the most promising method of tailings reprocessing will be underground leaching. There are a number of factors which are favourable for this process:

- The extremely low filtration coefficient of the pit edge cliff rocks ($1*10^{-2}$ - $1*10^{-3}$ m/day) in comparison with tailings (0,5-1,0 m/day), which means very low loss of working solution from the pit;
- The high tailings filtration coefficient makes it possible to increase significantly the distance between the feeding boreholes, and to ensure that the saturated solution is collected through the existing drainage drift of pit № 2, after drilling a special water collecting borehole for connection with the drainage drift;
- Leaching can be employed throughout the year if there is a 1.5-2.0m thick protective layer above the tailings;
- Secondary wastes remain in the same place after copper extraction, so no expenses are incurred for transportation or conservation.

As was the case at Visokogorsky OMPE, implementing this project means several tasks can be solved simultaneously:

- Work can continue at Gaiskiy OMPE without a new surface tailings storage facility building;
- Reclamation of Pit № 2 can be achieved without any additional investments;
- Stopping exploitation, and reclaiming the working OMPE tailings storage facility would significantly improve the ecological situation in the region of Gai city;
- As far as the fulfilment of the conservation recommendations (tailings preservation) for the tailings in pit № 2 are concerned, further technological development will make it possible to fulfil their secondary development and to extract the copper and other useful components they contain.

Conclusions

At present, the designed project and the results of its environmental impact assessment are awaiting consideration by the Orenburgskaya Oblast's environmental protection authorities.

After two cases of observation, it is possible to say that the usage of developed pits as tailings storage facilities for ore mining and processing plants is a very promising method of reclaiming these pits. This is a means of solving both the socio-economic and the environmental problems of mining regions.

As stated above, there are many developed open pits and large volumes of wastes containing useful substances in the Ural region.

Hence, the suggested approach can be considered a means of long-term sustainable development for the mining industry in the region, combining the desire to satisfy the region's development needs with preservation of the natural environment and reclamation of areas disturbed by mining activities. It is obvious that such an approach can also be successfully employed in other regions with a highly developed mining industry.

CHAPTER 28
Removal of zinc from iron ore materials prior to charging into blast furnace

Prof. Alexei Kapustin, Alexander Tomash

Priazovsky State Technical University
vul. Universytets'ka 7
Mariupol 87500
Ukraine

Summary

This paper discusses the use of a bishofit mineral for the removal of zinc compounds during preparation of metallurgical raw material. The technology differs with a simplicity and high efficiency and can be used in any metallurgical plant. The received product (zinc chloride) can be sold on the market.

Bishofit for the removal of zinc

Any ecological problem can and will be solved if doing so helps create a profit. This means that any contamination can be considered a potentially marketable product. For example, such highly dangerous substances as benzopyrene and dioxin, can, when properly cleaned, be sold at the prices listed below.

Table 1.

№	Substance	Price, €/g, Sigma, 2002
1	Benzopyrene	68
2	1,2,7,8-Dioxin	213
3	PbO	2
4	SO_2	1

Even such an irksome contaminant as an oxide of sulphur formed during burning of fuel in power stations can be realized on the market.

Furthermore, at any large plant, for example steel works, huge quantities of various compounds and chemical elements, which may be toxic contaminants are processed. Their transformation during the technical process is not monitored and there is no information about what happens to them at a later stage. Table 2 provides data on some of the chemical elements in a large metallurgical plant. When we are able trace the path and transformations of each element or connection, it is possible not only to prevent its emission into the environment, but also, by creating the appropriate technology, to obtain a marketable product.

One such technological development for the removal of Sb, As, Pb, and especially of zinc, is described in Table 2.

W. Leal Filho and I. Butorina (eds.),
Approaches to Handling Environmental Problems in the Mining and Metallurgical Regions, 289–293.
© 2003 *Kluwer Academic Publishers. Printed in the Netherlands.*

290

Element	Balance account
Fe	> 99,9 %
S	> 90 %
Hg	20/80 %
Tl	15
Au	8
Ag	1
Br	~ 0

Table 2

At the Ilyich Iron & Steel Works the mean content of zinc in the agglomerate structure at present amounts to 0,2 %. The blast process using raw material with an increased zinc content causes significant difficulties due to its harmful impact. Zinc dust resides in gas pipelines; zinc deposits are formed inside the furnace; and zinc penetrates into the refractory lining, destroying it and, in some cases, leading to the breakage of the blast furnace shell. The negative impact of zinc is aggravated by the fact that it accumulates in blast furnaces. Sublimated in the lower layers of the blast furnace, it is condensed in the top layers. The incoming zinc is added to the zinc already in circulation, and the quantity of zinc in the blast furnace is increased. The deposits in Venture pipes of the gas-cleaning system in blast furnaces containing more than 40 % zinc, testify to a significant accumulation of zinc in the working space of blast furnaces.

The negative effect of zinc on blast furnace operation means that zinc bearing sludge cannot be used. Their further participation in production process requires the development of an effective way of removing zinc at the stage of blast blend preparation.

The proposals for the pre-calcinating of zinc bearing sludge on the KM-14 conveyor machine in the limestone-calcinating plant with a view to removing the zinc, require large-scale renovation of the plant and considerable capital and operation costs. Sludges are characterized by their high moisture content, fine size and, hence, extremely low gas permeability. this makes it impossible to process them on the sinter plant, by sucking the air through the bulk layer without pre-pelletising of the sludge. The application of such technology will require not only renovation units for transferring the load to conveyor machines, but also the construction of a sludge pelletising plant with pelletising drums and granulators of a different design. Processing of sludge also involves increased consumption of solid fuel. Furthermore, the selectivity of the given process is very low.

The lowest costs, at high efficiency of zinc removal, can be achieved by introducing chloride components to a blend structure. Thus, during the sintering of the agglomerate zinc chloride $ZnCl_2$ is formed. The proposed technology is based on the reaction between zinc compounds and chlorides which is accompanied by the formation of volatile zinc chloride (II):

$Zn_2 + + 2\ Cl^-$ and $ZnCl_2$

The volatility of zinc chloride is given in Table 3.

Table 3

Temperature, °C	Partial pressure oh Zn, mm Hg
428	1
508	10
648	100
732	760

As follows from the Table all zinc chloride is sublimated at a temperature of 732 °C. As the temperature in the sintering zone exceeds 1300 °C, all zinc chloride is removed. Capital investment is not essential for implementation of this technology. The disadvantage of this method of zinc removal, which makes it unsuitable for widespread use, is the absence of the appropriate Cl-containing additive. Calcium chloride $CaCl_2$ recommended for this purpose, does not occur naturally in large quantities. Being by-product of the chemical industry, its price, at around 300 $ per т, is high, which makes its application economically ineffective. The use of cooking salt, NaCl, as an additive was not successful, because of the formation of an oxide Na_2O, which has a harmful effect on the blast process.

We propose adding bishofit $MgCL_2 * 6H_2O$ into sinter for zinc removal. Bishofit is a natural mineral. There are significant deposits of this mineral in the Volgograd basin. The deposits are estimated at 164,8 million tons. The percentage of the main substance, $MgCl_2 * 6H_2O$, in a skew field of the mineral exceeds 95 %. The mineral does not have a negative effect on human health or on the environment.

Bishofit solution can be delivered in tanks to sinter plant and added to the sinter blend. The cost of the natural solution is fairly low and mainly determined by transportation costs. The cost of 1 m^3 is approximately $50. Bishofit can be delivered, either in solution or in lumps, directly from the Volgograd deposit.

A concentrated solution of bishofit can be supplied to a pelletising drum together with water for pelletising. This method will ensure that it mixes uniformly with a blend. During sintering magnesium chloride will be subject to the following chemical reactions within the blend structure:

$$Mgcl_2 * 6H_2O = Mgcl_2 + 6H_2O$$

$$Mgcl_2 + ZnO = Zncl_2 + MgO$$

Zinc chloride will turn into gas inside the sintering zone and will be removed, together with sucked-through air. The oxide of magnesium will become part of the sinter composition. Sintering of a zinc-bearing blend with bishofit additives, in laboratory conditions, has shown that more than 99 % of zinc, lead and As (table 4) is removed.

Table 4

	Initial blend	Agglomerate	Agglomerate after bishofit
MgO	1.35 %	1.52 %	1.67
SiO$_2$	8.94 %	8.73 %	8.89
CaO	11.0 %	9.53 %	11.31
Mn	0.14 %	0.12 %	0.12
Fe	51.7 %	55.9 %	53.5
K	14 ppm	11 ppm	13 ppm
Na	12 ppm	11 ppm	13 ppm
Zn	0.09 %	0.085 %	7 ppm
Pb	0.18 %	0.181 %	2 ppm
As	0.08 %	0.075 %	0

Thus, the introduction of a magnesium chloride into a blend will make it possible to obtain a zinc-free product when using zinc bearing sludges. If there is an excess of bishofit in the blend after zinc removal, calcium chloride having high chemical activity to chlorine will form. CaCl$_2$ will remain in an agglomerate structure. In the blast furnace calcium chloride, falling into the blast slag, will influence its properties similarly to CaF$_2$, i.e. will increase its mobility, which will have a positive effect on the blast furnace operation. Formation of HCl will not take place in the presence of a large quantity of basic components. Therefore, the introduction of magnesium chloride into the blend will not cause an increase of toxic emissions to the atmosphere or a deterioration of sinter quality. When cooling the gas flow, the removed zinc chloride will accumulate in the collector and, to a certain extent, in filters, in the form of needle-shaped dust which does not form deposits. Together with iron dust, zinc chloride will be returned into the blend. It will be removed once more from the sintering zone.

Thus, zinc chloride will circulate in iron ore dust and sinter structure between the conveyor machine and the amalgamator drum, and will gradually accumulate. After several cycles the circulation of ZnCl$_2$ will be interrupted. The iron ore dust from manifold and filters will be collected periodically and sent to a chemical-metallurgical plant for the extraction of zinc chloride. The recycling of dust will include the dissolution of zinc chloride in water, evaporation and separation of zinc chloride of the grade "Technical". The dust from collector and filters which is not dissolved in water, has no significant commercial value and can be dumped.

Zinc chloride obtained as a result of the above process can be realized on the market. The cost of zinc chloride graded "technical" (50 % water solution), delivered to the European market, is USD 542-USD 580 per ton. The cost of zinc chloride recalculated for dry substance around USD 1.200 per ton.

Conclusions

With an annual output of agglomerate equal to 10 mln. t and a zinc content of about 0,2 %, a sinter plant takes in 20 thousand tons of zinc per year. In order to remove this, 67 thousand tons of bishofit or 111,5 thousand m^3 of solution are needed. The additional costs of purchase and delivery of solution may vary according to whether the solution is supplied directly from the deposit or transported. The cost of purchasing bishofit with consideration of replacing dolomitized limestone will also vary within the above estimates.

The cost of the obtained zinc chloride also allows a noticeable annual profit. Specific profit per ton of metal produced by the steel works is around USD 5,00. The calculation of the annual profit does not include losses caused by the failure of the refractory lining before time, prevention of deposit formation, cost of unscheduled downtime due to clogging of gas pipelines, or a decrease of specific consumption of coke for iron melting due to over consumption for sublimation or the recovery of zinc compounds in the working space of the blast furnace.

Thematic Index

W. Leal Filho and I. Butorina (eds.),
Approaches to Handling Environmental Problems in the Mining and Metallurgical Regions, 295–296.
© 2003 *Kluwer Academic Publishers. Printed in the Netherlands.*

296